Lecture Notes in Computer Science 12852

More information about this subseries at http://www.springer.com/series/7408

Ibrahim Habli · Mark Sujan ·
Friedemann Bitsch (Eds.)

Computer Safety, Reliability, and Security

40th International Conference, SAFECOMP 2021
York, UK, September 8–10, 2021
Proceedings

 Springer

Editors
Ibrahim Habli ⓘ
University of York
York, UK

Mark Sujan ⓘ
Human Factors Everywhere Ltd.
Woking, UK

Friedemann Bitsch ⓘ
Thales Deutschland GmbH
Ditzingen, Germany

ISSN 0302-9743   ISSN 1611-3349   (electronic)
Lecture Notes in Computer Science
ISBN 978-3-030-83902-4   ISBN 978-3-030-83903-1   (eBook)
https://doi.org/10.1007/978-3-030-83903-1

LNCS Sublibrary: SL2 – Programming and Software Engineering

This Springer imprint is published by the registered company Springer Nature Switzerland AG
The registered company address is: Gewerbestrasse 11, 6330 Cham, Switzerland

Preface

This volume (LNCS 12852) contains the proceedings of the 40th International Conference on Computer Safety, Reliability and Security (SAFECOMP 2021) held during September 8–10, 2021. Due to the continued COVID-19 pandemic, SAFECOMP 2021 took place as a hybrid event, offering both in-person presentations and limited attendance at the University of York, UK, in accordance with suggested precautions, as well as the opportunity to present and attend online. The conference series was established in 1979 by the European Workshop on Industrial Computer Systems, Technical Committee 7 on Reliability, Safety and Security (EWICS TC7). Since then, SAFE-COMP has contributed to progressing the state of the art of dependable computer systems and their application in safety-critical and security-critical systems. SAFE-COMP covers all areas of dependable systems, including embedded systems, cyber-physical systems, Internet of Things, systems-of-systems, cybersecurity, digital society, and many more. In recent years, autonomous systems, particularly those that incorporate machine learning models, have become increasingly important topics, and, in line with this development, the assurance of the safety and security of such systems in real-world applications is one of the highest and most challenging priorities. This is reflected in the keynote presentations as well as in the key theme of SAFECOMP 2021, which was "Safe Human – Robotic & Autonomous Systems Interaction".

The International Program Committee consisted of 51 members from 18 countries. The review process was thorough, single-blind (i.e., authors did not know the reviewers' identity), and each manuscript was reviewed by at least three independent reviewers. The merits of each paper were evaluated by the Program Committee members during a virtual meeting in April 2021. In total, after desk-rejecting papers that were beyond the scope of the conference or did not meet the essential formatting requitements, 76 submissions were peer-reviewed, and 17 manuscripts were selected for presentation and inclusion in the proceedings (an acceptance rate of 22%). We would like to thank all the reviewers and sub-reviewers for their contributions to ensuring an interesting and high-quality conference program.

We were pleased to host three stimulating keynote presentations. Prof. Adnan Darwiche (UCLA, USA) talked about "Empowering data with knowledge and reasoning". Prof. Neville Stanton (Southampton University, UK) gave a provocative presentation about "Driver reactions to autonomous vehicles". Prof. Sadie Creese (University of Oxford, UK) shared with the audience "Thoughts for a cybersecurity framework for protecting machine learning/AI systems".

As in previous years, SAFECOMP was organized as a single-track event to enable participants to attend all sessions, and to allow for networking during breaks and social events, both in person as well as via the electronic conference platform. The main conference was preceded by a day of topical workshops. This year, there were five workshops: 16th International Workshop on Dependable Smart Embedded Cyber-Physical Systems and Systems-of-Systems (DECSoS 2021); 2nd International

Workshop on Dependable Development-Operation Continuum Methods for Dependable Cyber-Physical Systems (DepDevOps 2021); 1st International Workshop on Multi-concern Assurance Practices in Software Design (MAPSOD 2021); 2nd International Workshop on Underpinnings for Safe Distributed AI (USDAI 2021); and 4th International Workshop on Artificial Intelligence Safety Engineering (WAISE 2021). The papers presented at these workshops are published in a separate LNCS volume (12853).

We would like to express our sincere gratitude to the many people whose contributions made SAFECOMP 2021 possible: the authors who submitted manuscripts; the invited keynote speakers; Prof. John McDermid as conference chair; the Program Committee members and external reviewers; EWICS TC7 and chair person Prof. Francesca Saglietti; the conference sponsors and supporting organisations; Friedemann Bitsch as the publications chair; Erwin Schoitsch and Simos Gerasimu as workshop co-chairs; Simon Burton as the industry chair; and the local Organization Committee members Sarah Heathwood, Dawn Forrester, and Alex King, who managed all of the practical arrangements and who ensured that the conference was an interesting experience for all.

We hope that readers will find these conference proceedings interesting and thought provoking.

September 2021

Ibrahim Habli
Mark Sujan

Organization

EWICS TC7 Chair

Francesca Saglietti   University of Erlangen-Nuremberg, Germany

General Chair

John McDermid   University of York, UK

Program Co-chairs

Ibrahim Habli   University of York, UK
Mark Sujan   Human Factors Everywhere, UK

General Workshop Co-chairs

Simos Gerasimou   University of York, UK
Erwin Schoitsch   AIT Austrian Institute of Technology, Austria

Publication Chair

Friedemann Bitsch   Thales Deutschland GmbH, Germany

Local Organizing Committee

Dawn Forrester   University of York, UK
Sarah Heathwood   University of York, UK
Alex King   University of York, UK

Industry Chair

Simon Burton   Fraunhofer IKS, Germany

International Program Committee

Uwe Becker   Draeger Medical GmbH, Germany
Peter G. Bishop   Adelard, UK
Friedemann Bitsch   Thales Deutschland GmbH, Germany
Sandro Bologna   Associazione Italiana Esperti Infrastrutture Critiche, Italy
Andrea Bondavalli   University of Florence, Italy
Jens Braband   Siemens AG, Germany

Elena Troubitsyna	KTH Royal Institute of Technology, Sweden
Marcel Verhoef	European Space Agency, The Netherlands
Marcus Völp	University of Luxembourg, Luxembourg
Hélène Waeselynck	LAAS-CNRS, France

Sub-reviewers

Victor Bandur	McMaster University, Canada
Jana Berger	RWTH Aachen University, Germany
Andrea Ceccarelli	University of Florence, Italy
Lorenzo De Donato	University of Naples Federico II, Italy
José M. Gaspar Sánchez	KTH Royal Institute of Technology, Sweden
Magnus Gyllenhammar	Zenseact, Sweden
Richard Hawkins	University of York, UK
Yassir Idmessaoud	LAAS-CNRS, France
Shahid Khan	RWTH Aachen University, Germany
Ryo Kurachi	Nagoya University, Japan
Stefano Marrone	University of Naples Federico II, Italy
Yutaka Matsuno	Nihon University, Japan
Roberto Nardone	Università Mediterranea Di Reggio Calabria, Italy
Mark Nicholson	University of York, UK
Thomas Noll	RWTH Aachen University, Germany
Vera Pantelic	McMaster University, Canada
Michael Parsons	University of York, UK
Colin Paterson	University of York, UK
Chiara Picardi	University of York, UK
Muhammad Rusyadi Ramli	KTH Royal Institute of Technology, Sweden
Jan Reich	Fraunhofer Institute for Experimental Software Engineering, Germany
Toru Sakon	CAV Technologies Co., Ltd., Japan
Mehdi Saman Azari	Linnaeus University, Sweden
Thomas Santen	TU Berlin, Germany
Andreas Schmidt	Fraunhofer Institute for Experimental Software Engineering IESE, Germany
Kaustubh Sridhar	University of Pennsylvania, USA
Lifei Tang	KTH Royal Institute of Technology, Sweden
Maryam Zahid	Mälardalen University, Sweden

Gold Sponsor

Intel

Supporting Institutions

European Workshop on
Industrial Computer Systems –
Reliability, Safety and Security

University of York

Assuring Autonomy International
Programme

Human Factors Everywhere Ltd

Austrian Institute of Technology

Thales Deutschland GmbH

Lecture Notes in Computer Science
(LNCS), Springer Nature

Chartered Institute of
Ergonomics & Human Factors

Chartered Institute
of Ergonomics
& Human Factors

European Training Network for
Safer Autonomous Systems

Safety-Critical Systems Club

European Network of Clubs for
Reliability and Safety of
Software-Intensive Systems

German Computer Society

Informationstechnische
Gesellschaft

Electronic Components
and Systems for European
Leadership - Austria

ARTEMIS Industry Association

Verband Österreichischer
Software Industrie

Austrian Computer Society

European Research Consortium for
Informatics and Mathematics

Contents

Machine Learning Applications

Safety Validation and Simulation

Fault Tolerance

Machine Learning Safety Assurance

DeepCert: Verification of Contextually Relevant Robustness for Neural Network Image Classifiers

Colin Paterson[1(✉)], Haoze Wu[2], John Grese[3], Radu Calinescu[1], Corina S. Păsăreanu[3], and Clark Barrett[2]

[1] University of York, York, UK
colin.paterson@york.ac.uk
[2] Stanford University, Stanford, USA
[3] Carnegie Mellon University, Silicon Valley, Pittsburgh, USA

Abstract. We introduce DeepCert, a tool-supported method for verifying the robustness of deep neural network (DNN) image classifiers to *contextually relevant perturbations* such as blur, haze, and changes in image contrast. While the robustness of DNN classifiers has been the subject of intense research in recent years, the solutions delivered by this research focus on verifying DNN robustness to small perturbations in the images being classified, with perturbation magnitude measured using established L_p norms. This is useful for identifying potential adversarial attacks on DNN image classifiers, but cannot verify DNN robustness to contextually relevant image perturbations, which are typically not small when expressed with L_p norms. DeepCert addresses this underexplored verification problem by supporting: (1) the encoding of real-world image perturbations; (2) the systematic evaluation of contextually relevant DNN robustness, using both testing and formal verification; (3) the generation of contextually relevant counterexamples; and, through these, (4) the selection of DNN image classifiers suitable for the operational context (i) envisaged when a potentially safety-critical system is designed, or (ii) observed by a deployed system. We demonstrate the effectiveness of DeepCert by showing how it can be used to verify the robustness of DNN image classifiers build for two benchmark datasets ('German Traffic Sign' and 'CIFAR-10') to multiple contextually relevant perturbations.

Keywords: Deep neural network robustness · Deep neural network verification · Contextually relevant image perturbations

1 Introduction

Deep neural network (DNN) image classifiers are increasingly being proposed for use in safety critical applications [6,15,19,24], where their accuracy is quoted as close to, or exceeding, that of human operators [3]. It has been shown, however, that when the inputs to the classifier are subjected to small perturbations, even

© Springer Nature Switzerland AG 2021
I. Habli et al. (Eds.): SAFECOMP 2021, LNCS 12852, pp. 3–17, 2021.
https://doi.org/10.1007/978-3-030-83903-1_5

highly accurate DNNs can produce erroneous results [8,9,30]. This has lead to intense research into verification techniques that check whether a DNN is robust to perturbations within a small distance from a given input, where this distance is measured using an L_p norm (e.g., the Euclidean norm for $p = 2$) [4,12,13, 20]. These techniques are particularly useful for identifying potential adversarial attacks on DNNs [8,14,17,18]. They are also useful when small changes in the DNN inputs correspond to meaningful changes in the real world, e.g., to changes in the speed and course of an aircraft for the ACAS Xu DNN verified in [12].

For DNN image classifiers, small L_p-norm image changes are not always meaningful. Changes that may be more meaningful for such DNNs (e.g., image blurring, hazing, variations in lighting conditions, and other natural phenomena) can also cause misclassifications, but are difficult to map to small pixel variations [10,16], and thus cannot be examined using traditional DNN verification techniques. What is needed for the comparison and selection of DNN image classifiers used in safety-critical systems is a *contextually relevant robustness verification* method capable of assessing the robustness of DNNs to these real-world phenomena [1,2,25,31]. Moreover, this verification needs to be performed at DNN level (i.e., across large datasets with image samples from all relevant classes) rather than for a single sample image.

The tool-supported DeepCert[1] method introduced in our paper addresses these needs by enabling:

1. The formal encoding of contextually relevant image perturbations at quantified perturbation levels $\epsilon \in [0, 1]$.
2. The verification of contextually relevant DNN robustness, to establish how the accuracy of a DNN degrades as the perturbation level ϵ increases. DeepCert can perform this verification using either test-based (fast but approximate) or formal verification (slow but providing formal guarantees).
3. The generation of contextually relevant counterexamples. These counterexamples provide engineers with visually meaningful information about the level of blur, haze, etc. at which DNN classifiers stop working correctly.
4. The selection of DNNs appropriate for the operational context (i) envisaged when a safety-critical system is designed, or (ii) observed by the deployed system during operation.

We organised the rest of the paper as follows. Section 2 describes our Deep-Cert verification method, explaining its encoding of contextual perturbations, and detailing how it can be instantiated to use test-based and formal verification. Section 3 presents the DeepCert implementation, and Sect. 4 describes the experiments we performed to evaluate it. Finally, Sect. 5 discusses related work, and Sect. 6 provides a summary and outlines future research directions.

---

[1] <u>Deep</u> neural network <u>C</u>ont<u>e</u>xtual <u>r</u>obus<u>t</u>ness.

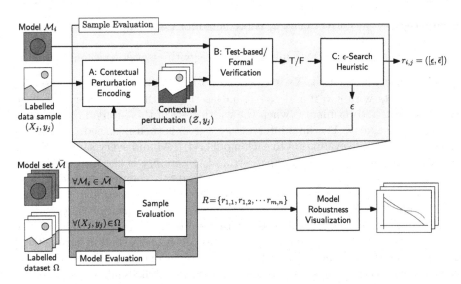

Fig. 1. DeepCert process for verifying contextually meaningful DNN robustness.

2   DeepCert Verification Method

2.1   Overview

Figure 1 shows our DeepCert method for the systematic verification of contextually relevant DNN robustness. DeepCert accepts as input a set of $m \geq 1$ DNN models, $\bar{\mathcal{M}}$, and a dataset of $n \geq 1$ labelled image samples, Ω. Each element $u \in \Omega$ is a tuple $u = (X, y)$ where $X \in \mathcal{X}$ is the input sample, $\mathcal{X}$ is the DNN input space, and y is a label indicating the class into which the models should place the sample. During model evaluation, each model $\mathcal{M}_i \in \bar{\mathcal{M}}$ is evaluated against each labelled data sample $(X_j, y_j) \in \Omega$, to find a robustness measure for that sample. The results are then presented to the engineer as visualisations that enable model-level contextual robustness evaluation and comparison.

The sample evaluation (top of Fig. 1) is a three-stage iterative process. The first stage (A) encodes the contextual perturbation using a function $g : \mathcal{X} \times [0,1] \rightarrow 2^{\mathcal{X}}$ that maps the data sample $X_j \in \mathcal{X}$ and a *perturbation level* $\epsilon \in [0,1]$ to a set of DNN inputs $\mathcal{Z} = g(X_j, \epsilon) \in 2^{\mathcal{X}}$ corresponding to images obtained by applying the contextual perturbation being verified (e.g., haze or blur) to the original image sample X_j. As we explain later in this section, g applies the perturbation at level ϵ when DeepCert employs test-based verification, and at *all* levels in the range $[0, \epsilon]$ when DeepCert employs formal verification.

The second stage (B) verifies whether the model $\mathcal{M}_i$ is robust to the contextual perturbation $(\mathcal{Z}, y_j)$, i.e., whether it classifies all images from $\mathcal{Z}$ as belonging to class y_j. The output of this stage is a Boolean value, true (T) or false (F).

The final state (C) is a search heuristic that supplies the ϵ value used for the contextul perturbation encoding from stage A, and employs binary search to identify perturbation level bounds $\underline{\epsilon}, \bar{\epsilon} \in [0, 1]$ such that:

– either $\underline{\epsilon} < \bar{\epsilon}$, the correct class y_j is predicted for $\epsilon = \underline{\epsilon}$, and a misclassification occurs for $\epsilon = \bar{\epsilon}$;
– or $\underline{\epsilon} = \bar{\epsilon} = 0$, and the DNN misclassifies X_j (with no perturbation applied).

After checking whether X_j is classified correctly by model $\mathcal{M}_i$, the search heuristic starts with $\underline{\epsilon} = 0$ and $\bar{\epsilon} = 1$, halves the width of the interval $[\underline{\epsilon}, \bar{\epsilon}]$ in each iteration, and terminates when the width $\bar{\epsilon} - \underline{\epsilon}$ of this interval drops below a predefined value ω. The final interval $r_{i,j} = [\underline{\epsilon}, \bar{\epsilon}]$ is then returned.

Applying sample evaluation to each model $\mathcal{M}_i \in \mathcal{M}$ and every sample $X_j \in \Omega$ provides a result set $R = \{r_{1,1}, r_{1,2}, \cdots r_{m,n}\}$, where $r_{i,j}$ is the interval for the i-th model and j-th image sample. For each result, a counterexample X'_j can be generated, if one exists (i.e., if $\underline{\epsilon} < 1$), by perturbing the sample X_j at level $\epsilon = \bar{\epsilon}$. Evaluating X'_j using model $\mathcal{M}_i$ produces a misclassification label $\hat{y}_j$.

Visualisations of model and class robustness are then produced in which the accuracy of the models is presented as a function of the perturbation parameter ϵ. By examining the accuracy of models across the range of expected perturbations, we can identify the conditions under which model switch should occur, e.g. one model may perform well at low levels of haze whilst a second may be superior as the level of haze present increases. Where the visualisations indicate that a particular class accuracy is highly sensitive to changes in ϵ this may indicate the need to choose a less sensitive model, or to gather additional training data.

2.2   DeepCert Instantiation for Test-Based Verification

For test-based verification, the contextual perturbation encoding function g maps an image X to a set $\mathcal{Z}$ comprising a single modified image X' obtained by applying a perturbation function:

$$x'_{i,j} = perturbation(X_{i,j}, \epsilon), \tag{1}$$

where $x'_{i,j}$ is the pixel at position (i, j) in the modified image X' and $X_{i,j}$ is a subset of pixels from the original image X. For colour images, a sample X is encoded as an array of pixels each of which is a 3-tuple of values representing the red, green and blue components of the colour in that pixel. DeepCert may utilize any perturbation which can be coded using (1) and three typical contextual perturbations are shown in (Fig. 2).

Haze Encoding. Haze represents a phenomenon where particles in the atmosphere scatter the light reaching the observer. The effect is to drain colour from the image and create a veil of white, or coloured, mist over the image. While realistic approaches to the modelling of haze require complex models [32], simplifying assumptions can be made. Assuming the haze is uniform, a haze colour may be defined as $C^f = (r, g, b)$ and applied to the image as:

$$x'_{i,j} = (1 - \epsilon)x_{i,j} + \epsilon\, C^f \tag{2}$$

where $\epsilon \in [0, 1]$ is a proxy for the density of the haze. When $\epsilon = 0$ the image is unaltered and when $\epsilon = 1$ the image is a single solid colour C^f. Multiplication and addition are applied to the pixel in an element-wise manner.

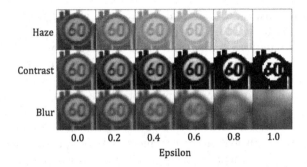

Fig. 2. Context perturbations applied to image sample

Contrast Variation Encoding. When fixed aperture lenses are employed, or when the dynamic range of the scene is extreme, the contrast in the image may become compressed. This effect may be modelled as:

$$x'_{i,j} = \text{Max}\left(0, \text{Min}\left(1, \frac{x_{i,j} - (0.5 * \epsilon)}{1 - \epsilon}\right)\right) \tag{3}$$

The effect of applying this function is to make bright parts of the image lighter and dark parts of the image darker.

Blur Encoding. Blurring in an image occurs when parts of the image are out of focus due to the limited capabilities of the optics employed in the system or when grease or water droplets are present on the lens. Blur can be synthesised using a convolutional kernel of size $2k_d + 1$ where the value of a pixel in the output image is calculated as a weighted sum of neighbouring pixels:

$$x'_{i,j} = \sum_{k=-k_d}^{k_d} \sum_{l=-k_d}^{k_d} \alpha_{k,l} \cdot x_{i+k,j+l} \tag{4}$$

The weights $\alpha_{k,l} \in (0,1)$ are calculated by discretising a two-dimensional Gaussian curve, where the sum of weights is equal to one, $\sum_{k=-k_d}^{k_d} \sum_{l=-k_d}^{k_d} \alpha_{k,l} = 1$. In our work, we define ϵ to be proportional to the standard deviation of the Gaussian distribution across the kernel and calculate the weights accordingly.

2.3 DeepCert Instantiation for Formal Verification

While test-based verification is computationally efficient, this efficiency is obtained by sacrificing completeness, i.e. if the perturbed image corresponding to an ϵ value of p is not an adversarial example, we cannot guarantee that the network is robust against all perturbations with ϵ smaller than p. Formal verification tools, by contrast, can provide such guarantees, but typically impose constraints on the types of models and perturbations which can be analysed.

To demonstrate the use of formal verification within DeepCert, we integrated it with Marabou [13], a complete verification toolbox for analyzing DNNs. Marabou handles common piecewise linear activation functions (e.g., ReLU, Max-Pool, Sign), integrates multiple state-of-the-art bound tightening techniques [21,26,28], and supports parallel processing [29]. Given a neural network and a verification query, Marabou constructs a set of linear and piecewise linear constraints. The satisfiability of the conjunction of those constraints is evaluated using either an MILP-solver or the Reluplex procedure [12]. Given sufficient time, Marabou will either conclude that the query is unsatisfiable or return a satisfying assignment to the query. For this work we extended Marbou to allow for the encoding of contextual perturbations using an input perturbation function, as detailed below for haze.

Haze Encoding. Given a DNN model $\mathcal{M}$, an image X, a fog colour C^f, and a maximum perturbation bound p, we introduce variables $\boldsymbol{X}, \boldsymbol{Y}$ and ϵ, denoting the DNN inputs, the DNN outputs and the perturbation bound, respectively. $\boldsymbol{X}$ has the same shape as X. We then construct the following set of constraints:

$$\boldsymbol{Y} = \mathcal{M}(\boldsymbol{X}) \tag{5a}$$

$$0 \le \epsilon \le p \tag{5b}$$

$$\bigwedge_{i \le |\boldsymbol{X}|} \left(\boldsymbol{x}_i = (1 - \epsilon)x_i + \epsilon\, C^f \right) \tag{5c}$$

$$\bigvee_{\substack{i <= |\boldsymbol{Y}| \\ \boldsymbol{y}_i \ne \boldsymbol{y}_{real}}} \boldsymbol{y}_i \ge \boldsymbol{y}_{real} \tag{5d}$$

Checking the satisfiability of the constraints allows us to state if the network is robust against the haze perturbation for $\epsilon \le p$. Constraint (5a) denotes the relationship between $\boldsymbol{X}$ and $\boldsymbol{Y}$. It is a piecewise linear constraint if $\mathcal{M}$ only contains piecewise linear activation functions. Constraint (5b) represents the perturbation bounds. Constraint (5c) defines the input variables as results of the hazing perturbation. Finally, let $\boldsymbol{y}_{real}$ be the correct label, constraint (5d) denotes that the output variable corresponding to the correct label is not greater than that of some other label. The network is locally adversarially robust against haze perturbation with $\epsilon \le p$ if, and only if, the conjunction of the constraints above is unsatisfiable. If the constraints above is satisfiable, there exists a perturbation within ϵ such that some output other than $\boldsymbol{y}_{real}$ is maximal.

3 Implementation

We implemented our method using a Python framework which we have made available on our tool website https://deepcert.github.io. The repository includes all models used in the paper, the code for the DeepCert tool with the encoded perturbations presented in the paper, the supporting scripts required to generate the performance visualisations and instructions on how to use the framework. In addition, a version of Marabou is provided with a Python interface in which the haze perturbation from the previous section is encoded.

Table 1. German Speed Sign Classification: Data and Models

(a) Data Sets

Class	Description	# Train	# Test
0	30 km/h	1980	720
1	50 km/h	2010	750
2	60 km/h	1260	450
3	70 km/h	1770	660
4	80 km/h	1650	630
5	100 km/h	1290	450
6	120 km/h	1260	450

(b) Models

Model	Description	Accuracy
1A	Small ReLu only model	0.816
1B		0.847
2A	Large ReLu only model	0.868
2B		0.866
3A	CNN Model	0.988
3B		0.984

4   Experimental Results

4.1   Case Study 1: Road Traffic Speed Sign Classification

Our first case study uses a subset of the German Traffic Sign benchmark [22] where each sample is a 32×32 RGB image. From this set we selected the seven classes which represented speed signs, the number of samples in each class are shown in Table 1a. We then built classification models at three levels of complexity with two models per level. The accuracy for all six models is reported in Table 1b which shows accuracy increasing with model complexity.

DeepCert with Test-Based Verification. For each model we applied our method using test-based verification, an initial value of $\epsilon = 0.5$ and a binary search heuristic with a maximum permissible interval of 0.002. Figure 3 shows the impact of haze on model accuracy as ϵ is increased. While Table 1b shows model 3A to be the most accurate (0.988) without perturbation, we note that for $\epsilon \gtrsim 0.7$, model 3B achieves superior accuracy. This behaviour is more clearly seen if we consider the ReLu-only models. Here model 2A has the best initial performance, but this rapidly deteriorates as ϵ increases such that other models are superior for even small amounts of haze.

These results demonstrate the dangers of selecting a model on the basis of the accuracy reported for unperturbed samples, and show how DeepCert enables a more meaningful model selection for the operational context. Indeed, were the system to be equipped with additional sensing, to assess the level of haze present, the engineer may choose to switch between models as the level of haze increased.

Our method also allows for the identification of those classes particularly susceptible to contextual perturbations. Figure 4 shows the performance of the convolutional neural network (CNN) models at different levels of perturbation. We note that class 1 is largely insensitive to haze, this is because an image perturbed with $\epsilon = 1$ results in a solid colour image which is classified as class 1 by both models. For all other classes the accuracy reduces as haze increases. The amount of degradation is seen to be dependent on the sample class and the model used. For example, class 0 is more robust to haze in model 3B than in 3A with class 3 more robust in model 3A.

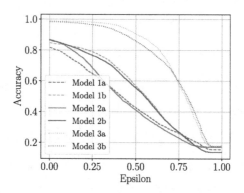

Fig. 3. Model robustness to haze.

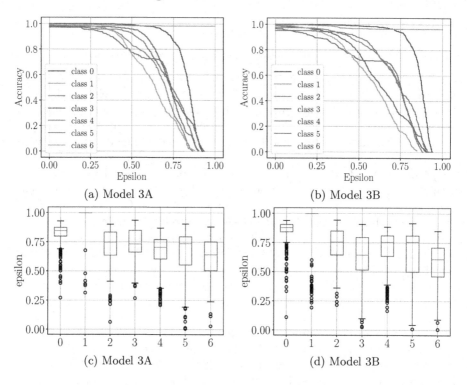

(a) Model 3A

(b) Model 3B

(c) Model 3A

(d) Model 3B

Fig. 4. Model Robustness with respect to haze

Figures 4c and 4d show the distribution of ϵ values required to cause misclassification where circles indicates samples identified as outliers. For class 3 we see that a number of samples are misclassified for small perturbations using model 3B but not 3A. An engineer wishing to deploy model 3B may examine these outliers to determine any correlation in image features. This may then allows for mitigation strategies at run-time or retraining with additional data samples.

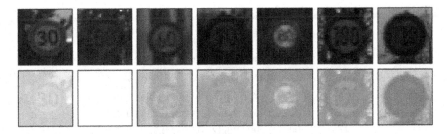

Fig. 5. Counterexamples for model 3A. Upper row is the original image, lower row has perturbation applied at the average level required for misclassification.

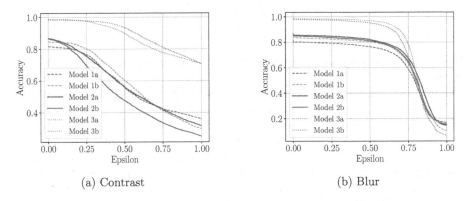

(a) Contrast   (b) Blur

Fig. 6. Model accuracy with respect to increased contrast and blur effects.

Our method also allows for the generation of meaningful counter examples for image based classifiers. Figure 5 shows counterexamples for model 3A and illustrates the average level of haze which each class can withstand before misclassification occurs. This visual representation of perturbation levels allows domain experts to consider the robustness of the model with respect to normal operating conditions.

Having demonstrated our approach using the haze perturbation we now show results for the contrast and blur effects. Model accuracy in the presence of these perturbations is shown in Fig. 6. We see that whilst the accuracy of models degrades as the amount of perturbation increases, the shape of the curves and the effect on individual models is different.

Model 3A was the most accurate model for much of the perturbation range under the effects of haze, while model 3B is superior with respect to contrast effects. We also see that while model 2B was relatively robust to haze, its robustness to contrast is poor. This shows that selecting a single model for all environmental conditions is unlikely to provide optimal performance. Our method allows for a greater understanding of models weaknesses when in the presence of natural phenomena and may allow for more intelligent choices to be made.

Table 2. Minimum ϵ values for l_∞ and hazing perturbation on test images.

Sample	Model 1A			Model 1B		
	Verification		Test	Verification		Test
	l_∞	Haze	Haze	l_∞	Haze	Haze
4	0.002	0.623	0.623	0.006	0.525	0.525
114	0.002	0.451	0.451	0.002	0.225	0.225
47	0.006	0.592	0.592	0.006	0.752	0.752
52	0.006	0.830	0.830	0.010	0.654	0.654
3	0.010	0.764	0.764	0.010	0.713	0.713
15	0.010	0.760	0.760	0.010	0.810	0.810

DeepCert with Formal Verification. For model 1A and 1B, we ran our method on the first 30 images correctly classified as class 3 in the test sets to compute the minimum ϵ values for hazing using contextual perturbations and for traditional l_∞ norm perturbations. For all 30 samples the value of ϵ found through formal verification was the same as that for the test based verification, although we can not guarantee this to be true for all samples in the testing set.

Table 2 shows selected results from the formal verification compared with the test-based verification Sample #4 has an l_∞ norm for model 1A that is lower than that of model 1B. This would indicate that model 1B is more robust. Examining contextual robustness, however, we see that model 1A is able to withstand more haze before misclassification occurs. A similar result is shown for sample #52. This time however model 1A would be judged more robust by the l_∞ measure whilst model 1B is more robust according to the contextual measure. Other samples report identical l_∞ measures between models (samples 114, 47, 3 and 15) yet their response to haze is different e.g. sample #114 using model 1A is able to withstand almost twice as much haze as model 1B.

These results demonstrate that our methods are able to use formal verification techniques, where the model form allows for such analysis. We also note that non-contextual point robustness is insufficient to assess the robustness of models in the presence of contextual perturbations.

4.2 Case Study 2: CIFAR-10

In order to demonstrate that our approach is applicable to a range of problems we applied our method to a second well known classification problem, CIFAR-10. The data set consists of 60,000 32×32 colour images in 10 classes with 5000 training images and 1000 test images per class. Table 3 shows the names of the classes in this benchmark. The complexity and diversity of the images in this set is a more challenging classification task than the traffic sign problem. We again constructed models of increasing complexity with two models at each level. The accuracy of these models for the unperturbed test set is given in Table 4.

Table 3. CIFAR-10 class descriptions

class	0	1	2	3	4	5	6	7	8	9
name	airplane	automobile	bird	cat	deer	dog	frog	horse	ship	truck

Table 4. CIFAR-10 model accuracy

Model		Accuracy	Model		Accuracy	Model		Accuracy
4A	Small Relu	49.11	5A	Large Relu	53.20	6A	CNN	84.07
4B		47.45	5B		53.04	6B		85.17

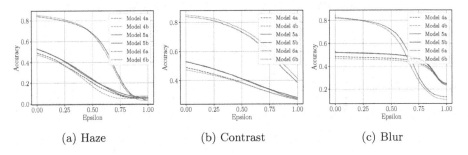

(a) Haze   (b) Contrast   (c) Blur

Fig. 7. CIFAR-10 model robustness

DeepCert with Test-Based Verification. Model accuracy in the presence of the three forms of contextual perturbation are shown in Fig. 7. We once more note the accuracy degrades as ϵ is increased for all perturbation types. For haze we observe a point at which the best model changes. This indicates that a system which is able to switch between models as the level of haze increases may demonstrate improved robustness. We also note that the CNN models outperform the simpler models by a significant margin under most conditions. For blur, however, when $\epsilon > 0.7$ the CNN models under perform the simpler models.

Figure 8 shows the class accuracy for the CNN models subjected to the blur perturbation. We observe that the performance of classes between the models varies as shown in the traffic sign sign study. The accuracy of class 3 in model 6A, for example, is lower than that seen in Model 6B until $\epsilon > 0.7$.

DeepCert with Formal Verification. Formal verification was applied to models 4A and 4B by again choosing 30 samples which we perturbed with haze. The results were in line with those found for the traffic sign model, but in addition we found a sample (#14) for model 4A which returned a lower robustness bound than when using test-based verification. Table 5 shows the predicted class $\hat{y}$ for this sample as ϵ is increased. We note that the sample is misclassified at $\epsilon = 0.0723$ which was found using Marabou, it then returns to classifying the sample correctly before misclassifying again at $\epsilon = 0.365$, the value found through testing. This confirms that, whilst testing may correctly identify the robustness bound for the majority of cases, formal verification is required for guarantees of robustness.

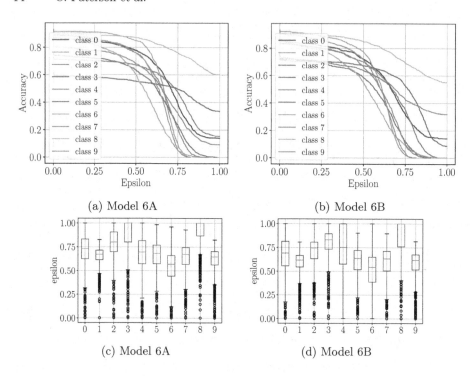

(a) Model 6A

(b) Model 6B

(c) Model 6A

(d) Model 6B

Fig. 8. CIFAR-10 class robustness with respect to blur

Table 5. Formal versus test-based verification, correct label $y = 9$

ϵ	$\hat{y}$	ϵ	$\hat{y}$
0.002	9	0.15	1
0.035	9	0.18	9
0.050	9	0.2	9
0.0723	1	0.03	9
0.1	1	0.365	2

5   Related Work

It is well known [23] that neural networks, including highly trained and smooth networks, are vulnerable to adversarial perturbations; these are small changes to an input (which are imperceptible to the human eye) that lead to mis-classifications. The vast majority of the work in this area focuses on formulating adversarial examples with respect to perturbations defined with L_p norms. The problem is typically formulated as follows: for a given network F and an input x, find an input x' for which $F(x') \neq F(x)$ while minimising $\|x - x'\|$.

The metric used to compute the distance between points is typically the Euclidean distance (L_2 norm), the Manhattan distance (L_1 norm), or the

Chebyshev distance (L_∞ norm). Methods for finding adversarial examples and for checking robustness of neural networks to adversarial perturbations range from heuristic and optimisation-based techniques [2,8,14,17,18] to formal analysis techniques which are based on constraint solving, interval analysis or abstract interpretation [4,5,7,11,12,27,28]. In contrast to these works, which focus on local robustness, we take a more global view, as we aim to evaluate models on many input points and use the results to assess and compare models and inform developers' choices. Furthermore, we aim to study more natural (contextual) perturbations, as we do not limit ourselves to L_p norms.

Other researchers have started to look into robustness verification beyond the L_p-norm threat model. For instance, Semantify-NN [16] addresses robustness verification against *semantic* adversarial attacks, such as colour shifting and lighting adjustment. It works by inserting semantic perturbation layers to the input layer of a given model, and leverages existing L_p-norm based verification tools to verify the model robustness against semantic perturbations. In our work, we also leverage an off-the-shelf verification tool (namely Marabou) to enable verification with respect to semantically meaningful perturbations. We do not modify the models, but instead encode the checks as Marabou queries.

6   Conclusions and Future Work

In this paper we have introduced DeepCert, a tool-supported method for the systematic verification of contextually relevant robustness for neural network classifiers. We have shown that the accuracy of a DNN image classifier is a function of the perturbation type to which sample images are exposed, and that through a systematic verification of the robustness with respect to these perturbations a more informed decision may be made to select a DNN model.

In future work we plan to investigate the use of alternative formal verification techniques with DeepCert, and the use of more complex models of natural phenomena, parameterised for use within the framework. We also intend to investigate methods to allow for the systematic assessment of robustness within regions of the input space e.g. rain drops on a lens affecting part of an image.

Acknowledgements. This research has received funding from the Assuring Autonomy International Programme project 'Assurance of Deep-Learning AI Techniques' and the UKRI project EP/V026747/1 'Trustworthy Autonomous Systems Node in Resilience'.

References

1. Ashmore, R., Calinescu, R., Paterson, C.: Assuring the machine learning lifecycle: desiderata, methods, and challenges (2019). arXiv preprint arXiv:1905.04223
2. Carlini, N., Wagner, D.: Towards evaluating the robustness of neural networks. In: 2017 IEEE Symposium on Security and Privacy, pp. 39–57. IEEE (2017)
3. De Fauw, J., et al.: Clinically applicable deep learning for diagnosis and referral in retinal disease. Nat. Med. **24**(9), 1342–1350 (2018)

4. Dutta, S., Jha, S., Sankaranarayanan, S., Tiwari, A.: Output range analysis for deep feedforward neural networks. In: NASA Formal Methods Symposium, pp. 121–138. Springer (2018)

5. Fischetti, M., Jo, J.: Deep Neural Networks as 0–1 mixed integer linear programs: a feasibility study (2017). arXiv preprint arXiv:1712.06174

6. Gauerhof, L., Hawkins, R., Picardi, C., Paterson, C., Hagiwara, Y., Habli, I.: Assuring the safety of machine learning for pedestrian detection at crossings. In: Casimiro, A., Ortmeier, F., Bitsch, F., Ferreira, P. (eds.) SAFECOMP 2020. LNCS, vol. 12234, pp. 197–212. Springer, Cham (2020). https://doi.org/10.1007/978-3-030-54549-9_13

7. Gehr, T., Mirman, M., Drachsler-Cohen, D., Tsankov, P., Chaudhuri, S., Vechev, M.T.: AI2: safety and robustness certification of neural networks with abstract interpretation. In: 2018 IEEE Symposium on Security and Privacy, pp. 3–18 (2018)

8. Goodfellow, I.J., Shlens, J., Szegedy, C.: Explaining and harnessing adversarial examples (2014). arXiv preprint arXiv:1412.6572

9. Grosse, K., Manoharan, P., Papernot, N., Backes, M., McDaniel, P.: On the (statistical) detection of adversarial examples (2017). arXiv preprint arXiv:1503.02531

10. Hamdi, A., Ghanem, B.: Towards analyzing semantic robustness of deep neural networks. In: European Conference on Computer Vision, pp. 22–38. Springer (2020)

11. Huang, X., Kwiatkowska, M., Wang, S., Wu, M.: Safety verification of deep neural networks. In: Majumdar, R., Kunčak, V. (eds.) CAV 2017. LNCS, vol. 10426, pp. 3–29. Springer, Cham (2017). https://doi.org/10.1007/978-3-319-63387-9_1

12. Katz, G., Barrett, C., Dill, D.L., Julian, K., Kochenderfer, M.J.: Reluplex: An efficient SMT solver for verifying deep neural networks. In: International Conference on Computer Aided Verification, pp. 97–117. Springer (2017)

13. Katz, G., et al.: The marabou framework for verification and analysis of deep neural networks. In: Dillig, I., Tasiran, S. (eds.) CAV 2019. LNCS, vol. 11561, pp. 443–452. Springer, Cham (2019). https://doi.org/10.1007/978-3-030-25540-4_26

14. Kurakin, A., Goodfellow, I., Bengio, S.: Adversarial machine learning at scale (2016). arXiv preprint arXiv:1611.01236

15. Mitani, A., et al.: Detection of anaemia from retinal fundus images via deep learning. Nat. Biomed. Eng. **4**(1), 18–27 (2020)

16. Mohapatra, J., Weng, T.W., Chen, P.Y., Liu, S., Daniel, L.: Towards verifying robustness of neural networks against a family of semantic perturbations. In: Proceedings of the IEEE/CVF Conference on Computer Vision and Pattern Recognition, pp. 244–252 (2020)

17. Moosavi-Dezfooli, S.M., Fawzi, A., Frossard, P.: Deepfool: a simple and accurate method to fool deep neural networks. In: Proceedings of the IEEE Conference on Computer Vision and Pattern Recognition, pp. 2574–2582 (2016)

18. Papernot, N., McDaniel, P., Jha, S., Fredrikson, M., Celik, Z.B., Swami, A.: The limitations of deep learning in adversarial settings. In: 2016 IEEE European Symposium on Security and Privacy, pp. 372–387. IEEE (2016)

19. Picardi, C., Paterson, C., Hawkins, R.D., Calinescu, R., Habli, I.: Assurance argument patterns and processes for machine learning in safety-related systems. In: Workshop on Artificial Intelligence Safety, pp. 23–30 (2020)

20. Pulina, L., Tacchella, A.: An abstraction-refinement approach to verification of artificial neural networks. In: CAV, pp. 243–257 (2010)

21. Singh, G., Gehr, T., Püschel, M., Vechev, M.: An abstract domain for certifying neural networks. Proc. ACM Program. Lang. **3**, 1–30 (2019)

22. Stallkamp, J., Schlipsing, M., Salmen, J., Igel, C.: The german traffic sign recognition benchmark: a multi-class classification competition. In: The 2011 International Joint Conference on Neural Networks, pp. 1453–1460. IEEE (2011)
23. Szegedy, C., et al.: Intriguing properties of neural networks (2013). arXiv:1312.6199
24. Tabernik, D., Skočaj, D.: Deep learning for large-scale traffic-sign detection and recognition. IEEE Trans. Intell. Transp. Syst. **21**(4), 1427–1440 (2019)
25. Tian, Y., Pei, K., Jana, S., Ray, B.: Deeptest: automated testing of deep-neural-network-driven autonomous cars. In: Proceedings of the 40th International Conference on Software Engineering, pp. 303–314 (2018)
26. Tjeng, V., Xiao, K., Tedrake, R.: Evaluating robustness of neural networks with mixed integer programming (2017). arXiv preprint arXiv:1711.07356
27. Wang, S., Pei, K., Whitehouse, J., Yang, J., Jana, S.: Efficient formal safety analysis of neural networks. In: Advances in Neural Information Processing Systems 31: Annual Conference on Neural Information Processing Systems (2018)
28. Wang, S., Pei, K., Whitehouse, J., Yang, J., Jana, S.: Formal security analysis of neural networks using symbolic intervals. In: 27th USENIX Security Symposium (2018)
29. Wu, H., et al.: Parallelization techniques for verifying neural networks. In: 2020 Formal Methods in Computer Aided Design, pp. 128–137 (2020)
30. Yuan, X., He, P., Zhu, Q., Li, X.: Adversarial examples: attacks and defenses for deep learning. IEEE Trans. Neural Netw. Learn. Syst. **30**(9), 2805–2824 (2019)
31. Zhang, M., Zhang, Y., Zhang, L., Liu, C., Khurshid, S.: DeepRoad: GAN-based metamorphic testing and input validation framework for autonomous driving systems. In: 2018 33rd IEEE/ACM International Conference on Automated Software Engineering, pp. 132–142. IEEE (2018)
32. Zhang, N., Zhang, L., Cheng, Z.: Towards simulating foggy and hazy images and evaluating their authenticity. In: Liu, D., Xie, S., Li, Y., Zhao, D., El-Alfy, E.S., et al. (eds.) Neural Information Processing, pp. 405–415. Springer, Cham, USA (2017). https://doi.org/10.1007/978-3-319-70090-8_42

Could We Relieve AI/ML Models of the Responsibility of Providing Dependable Uncertainty Estimates? A Study on Outside-Model Uncertainty Estimates

Lisa Jöckel[✉] and Michael Kläs

Fraunhofer Institute for Experimental Software Engineering IESE, Fraunhofer Platz 1,
67663 Kaiserslautern, Germany
{lisa.joeckel,michael.klaes}@iese.fraunhofer.de

Abstract. Improvements in Artificial Intelligence (AI), especially in the area of neural networks, have led to calls to use them also in the context of safety-critical systems. However, current AI-based models are data-driven, so we cannot assure that they will provide the intended outcome for any input. To obtain information about the uncertainty remaining in their outcome, uncertainty estimation capabilities can be integrated during model building. However, the approach of providing accurate outcomes and dependable uncertainty estimates using the same model has limitations. Among others, estimates of such 'in-model' approaches are provided without statistical confidence, tend to be overconfident if not calibrated, and are hard to interpret and review by domain experts. An alternative 'outside-model' approach is the use of model-agnostic uncertainty wrappers (UWs). To investigate how well they perform in comparison to in-model approaches, we benchmarked them against deep ensembles, which can be considered the gold standard for in-model uncertainty estimation, as well as to the softmax outputs of a deep neural network as a baseline. Despite a slightly higher Brier score, the UW provides other benefits that are important in a safety-critical context, like considering a statistical confidence level and providing explainable uncertainty estimates through a decision tree considering human-interpretable semantic factors. Furthermore, in-model uncertainty estimates can be forwarded into an UW, combining advantages of both approaches.

Keywords: Uncertainty wrapper · Data-driven model · Machine learning · Benchmarking study · Traffic sign recognition · Automated driving · Deep ensemble · Uncertainty calibration · Uncertainty quantification

1 Introduction

The use of Machine Learning (ML) and other Artificial Intelligence (AI) approaches can provide solutions for tasks that are difficult to tackle with traditional software development approaches. In recent years, in particular, deep neural networks have massively improved the performance of various tasks related to perception and understanding [1,2].

© Springer Nature Switzerland AG 2021
I. Habli et al. (Eds.): SAFECOMP 2021, LNCS 12852, pp. 18–33, 2021.
https://doi.org/10.1007/978-3-030-83903-1_2

Thus, it is not surprising that there is an intention to use modern ML approaches also in the context of systems with safety requirements as they promise new or massively improved functionalities. However, it is still open how best to deal with components that rely on data-driven models (DDMs) such as deep neural networks, in a safety context considering their hard-to-predict behavior.

One major issue is that we can neither assume nor demonstrate that such data-driven components will provide the "correct" outcome for any input. Contrary to traditional components, the behavior of data-driven components is specified by example data and thus we need to live with a certain degree of uncertainty when using them.

Although uncertainty estimation is an active research area in AI, many DDMs do not provide dependable uncertainty estimates. Research in this area is dominated by benchmarks and proposals for ML methods that lead to DDMs that provide uncertainty estimates together with their main outcome. Yet, we see – especially from a safety perspective – limitations with these kinds of 'in-model' uncertainty estimates [3]. In particular, they violate the 'separation of concerns' principle [4] since they make the DDM itself responsible for providing dependable assessments of its performance.

In the best case, this leads to black-box uncertainty estimates that cannot be checked for plausibility by experts. Commonly, however, such estimates also ignore features that would help to provide better uncertainty estimates because these features do not improve the accuracy of the primary model outcome. For example, knowledge about the amount of precipitation does not help to decide whether a camera picture shows a traffic sign of type A or B, but it helps to assess the uncertainty in the provided result. In the worst case, the bad practice of providing uncertainty estimates that are either not calibrated or calibrated on the data used to train the model can make these estimates systematically overconfident.

To provide an alternative to existing 'in-model' uncertainty estimation approaches, Kläs and Sembach proposed the 'uncertainty wrapper' (UW) concept as an 'outside-model' approach [5]. The concept of model-agnostic UWs does not only avoid the above issues, but also allows addressing all three types of uncertainty sources considered in the onion shell model [6], namely model fit, data quality, and scope compliance. Furthermore, a confidence level can be set for the provided uncertainty estimates, which we consider essential for the use in a safety-critical application.

In previous work, Kläs and Jöckel illustrated how UWs can be applied in the context of pedestrian detection [7]. Compared to a naïve baseline approach, the UW provided more dependable uncertainty estimates improving all three components of the Brier score, a common measure for the quality of probabilistic predictions [8, 9].

Although the 'outside-model' uncertainty estimates of UWs have advantages from a conceptual perspective, it is still open how well UWs perform with respect to estimation quality in comparison to state-of-the-art 'in-model' approaches. Moreover, it is open how well UWs deal with limitations commonly observed in practices and whether it is possible to leverage synergies between 'in-model' and 'outside-model' approaches. To address these questions, this paper presents an experimental study in which DDMs with and without an UW were benchmarked under different settings.

The remainder of this paper is structured as follows: Sect. 2 gives an overview of related work on uncertainty estimation including some background on the concept of

UWs. Section 3 elaborates the addressed research questions and introduces the study design and the benchmarking metrics we used. Section 4 presents and discusses the study results, and Sect. 5 concludes the paper.

2   Related Work on Uncertainty Predictions

Uncertainty is a topic of increasing relevance in the field of ML. The objective is to better understand the sources of uncertainty and provide dependable information on how much we can rely on an outcome given by a specific DDM.

How the concept of uncertainty as discussed in ML is related to safety standards, such as IEC 61508, and the application rule VDE-AR-E 2842-61 is detailed by Kläs et al. in their recent work [10], in which they also illustrate how the UW pattern can help to handle uncertainty in compliance with safety constraints.

Sources of Uncertainty.   A number of classifications have been proposed for potential sources of uncertainty. The best-known is probably the distinction between aleatoric and epistemic uncertainty. In general, *aleatoric* uncertainty means uncertainty due to randomness, which is not systematic but rather unavoidable "noise". *Epistemic* uncertainty, on the other hand, means uncertainty that is systemic in the way it refers to phenomena that could be known in principle but are not considered [11]. The idea is that we need to accept aleatoric uncertainty but should try to reduce epistemic uncertainty, e.g., by collecting more and better data. In a concrete setting, however, the distinction which part of the uncertainty is aleatoric and which is epistemic depends on the viewpoint [11]. Moreover, quantifications are usually not comparable between modeling approaches due to different interpretations and hypothesis spaces.

A classification that is orthogonal to aleatoric and epistemic is proposed by the *onion shell model* [6], which distinguishes between model fit, input quality, and scope compliance uncertainty. It allows mathematically separating uncertainty attributed to (a) limitations in the DDM, (b) differences in the quality of the model input, and (c) the possibility that the model is applied outside its application scope [5]. Type (a) can be reduced by improving the DDM and can be measured by traditional model testing and performance metrics (e.g., mean absolute error or true positive rate). Type (b) can be tackled by modeling the influence of input quality on the quality of the DDM outcomes (e.g., how the performance changes given a certain input quality). Type (c) relies on defining the application scope and monitoring compliance (e.g., how similar the current situation is to the situations considered during model training and testing).

Outside-Model Estimations.   Building upon this classification, an 'uncertainty wrapper' (UW) framework has been proposed [7]. UWs provide 'outside-model' estimations for uncertainty following the separation of concerns principle. They are model-agnostic and consider the DDM they encapsulate as a black box (cf. Fig. 1). Factors that may influence input quality and scope compliance uncertainty are modeled in a quality and a scope model, respectively. In the context of traffic sign recognition (TSR), e.g., the obstruction of vision by rain or fog would represent quality factors and the GPS coordinates could indicate as a scope factor whether the model is being applied outside its target application scope (e.g., a specific country). Using a decision-tree-based approach, the quality impact model decomposes the target application scope into areas with similar uncertainties based on the quality factors and safeguarded with a statistical confidence level that can be freely chosen. The scope compliance model uses the data provided by the scope factors to calculate the probability that the DDM is applied outside its target application scope. This can include checking some fixed boundaries as well as calculations on similarity between the current input and the inputs considered during model development.

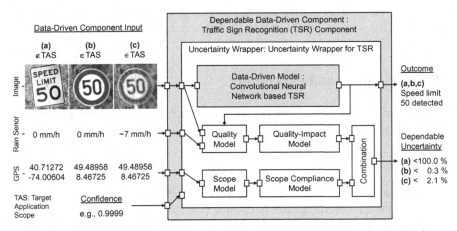

Fig. 1. Uncertainty wrapper architecture together with sample inputs and outputs [12].

In-Model Estimations. To date, the more established way to realize uncertainty estimation is to design the DDM itself such that it returns not only its categorical or binary outcome, but also the probability p that the provided outcome is correct. In the following, we will refer to approaches following this pattern as *'in-model'* approach.

Understanding *uncertainty* as the likelihood that the DDM outcome is not correct, the most naïve approach to estimate uncertainty is to determine the *overall error rate* on a sample as the global uncertainty estimate for all outcomes. Obviously, this estimation approach considers only model fit uncertainty.

There are classes of DDMs (e.g., logistic regression, Naïve Bayes, support vector machines) that provide by default a *preference value* between zero and one in addition to their outcome. Although these values are commonly interpreted as uncertainty estimates, there are limitations: Preference values do commonly nor represent real probabilities and are determined on training data, which favors overconfidence [3].

To address these limitations, *calibration methods* such as isotonic regression and Platt scaling [13] are applied as a kind of post-processing on the preference values. This, however, cannot solve the limitation that information sources are ignored that are only relevant for the uncertainty but not for the outcome.

Out-of-distribution and *novelty detection* methods [14] such as SafeML [15] provide means to detect whether a DDM is applied outside its intended application context.

However, they are also limited to scope-compliance-related uncertainty, which on the other hand is largely ignored by the 'in-model' approaches discussed above.

In the context of deep neural networks, which are our focus, Bayesian neural networks [16] and deep ensembles [17,16] are commonly proposed to provide uncertainty estimates. Since both approaches are computationally expensive, they are often approximated by using Monte Carlo Dropout [16]. Benchmarks and comparisons of state-of-the-art approaches for 'in-model' uncertainty estimation indicate that deep ensembles are currently the gold standard for neural networks considering estimation performance [18,19]. However, their computational demands increase linearly with the number of ensemble members during training and operation, which commonly challenges current hardware [16,18].

3   Study Planning and Execution

This chapter introduces and concretizes the addressed research questions, presents the derived study design, and explains the study execution.

3.1   Research Questions

A key question regarding the use of outside-model uncertainty estimation approaches is whether such model-agnostic approaches, which make no assumption about the internals of the encapsulated DDM, can achieve an estimation performance comparable to in-model uncertainty estimation approaches. In particular, we want to compare UW-based estimates with estimates based on existing in-model uncertainty estimation approaches that are either state-of-the-practice or state-of-the-art:

> *RQ1:* How does the uncertainty estimation performance of UWs and DDMs differ when trained and calibrated on data with sufficient examples of quality deficits?

Instead of considering in-model and outside-model approaches as competitors, we can also think about means to leverage synergies between the two kinds of approaches:

> *RQ2:* Can the uncertainty estimation performance of an UW be improved by also considering the uncertainty estimates of the DDM as a factor, besides other factors?

In practice, we commonly need to deal with imperfections. Thus, we investigate how the UW approach performs under conditions that can occur in real-world applications:

> *What are the implications if...*
> *RQ3.1:* no uncertainty factors with semantic meaning can be defined?
> *RQ3.2:* the DDM is not calibrated?
> *RQ3.3:* the DDM is trained on data that insufficiently covers relevant quality deficits?
> RQ3.4: a state-of-the-art DDM is not used due to limited computational resources?

Answering these questions can provide important insights for practical applications. RQ3.1 is linked to the question of whether it may be reasonable to build an UW even if no information sources on uncertainty other than the DDM are available. Answering RQ3.2 can help to decide whether a DDM has to be calibrated if we want to consider its uncertainty estimates as a factor in an UW. RQ3.3 investigates possible benefits of encapsulating a DDM that has blind spots on the quality issues it will face during its usage. Finally, RQ3.4 may help to identify cases where we can substitute a deep ensemble, i.e., the gold standard, with a more resource-efficient DDM.

3.2 Study Design and Variation Points

This section describes the context, design, and decisions in our study execution plan as summarized in Fig. 2 and detailed in Sect. 3.3. First, we will introduce the task of the DDM and the target application scope to which we refer in our evaluation. Next, we will describe the datasets used in the study and the augmentation used to enrich available datasets with quality deficits.

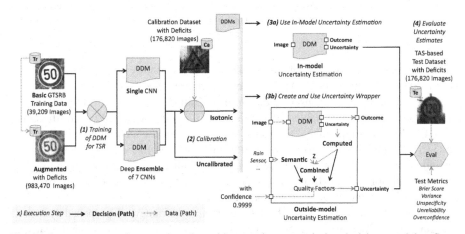

Fig. 2. Summary on study execution plan with execution steps, design decisions, and data flow.

Motivated by our research questions Table 1, summarizes relevant variation points in the study design, including the respective decisions and planned comparisons.

Task and Target Application Scope (TAS). We investigated our research questions using the example of traffic sign recognition, where the main task of the DDM is to correctly classify traffic signs on given images. The assumed target application scope was a roadworthy passenger car traveling in Germany at different points in time and faced with different weather conditions and related operation conditions, such as dirt on the camera lens. Note that our interest in the study is not on the performance of the DDM with respect to its primary outcome but the provided uncertainty estimate.

Datasets. To address our research questions, we need different kinds of data: *training data* to build the DDM and to identify situations that differ in their degree of uncertainty; *calibration data* to derive unbiased uncertainty estimators on unseen data; and

Table 1. Overview on the decisions and planned comparisons for each research question.

Research question	Training data	DDM	Calibration approach	UW quality factors	Comparison
RQ1	Augmented	Single + Ensemble	Isotonic	Semantic	*In-model vs. UW*
RQ2	Augmented	Single + Ensemble	Isotonic	*Combined*	vs. RQ1
RQ3.1	Augmented	Single + Ensemble	Isotonic	*Computed*	vs. RQ1 & RQ2
RQ3.2	Augmented	Single + Ensemble	*Uncalibrated*	Combined + Computed	vs. RQ2 & RQ3.1
RQ3.3	*Basic*	Single + Ensemble	Isotonic + Uncalibrated	All	vs. RQ1 to RQ3.2
RQ3.4	Augmented	*Single + Ensemble*	Isotonic + Uncalibrated	Semantic	Single vs. Ensemble

representative *test data* to evaluate the quality of the uncertainty estimates provided by the investigated in-model and outside-model uncertainty approaches.

The foundation of the datasets is the German Traffic Sign Recognition Benchmark (GTSRB) dataset [20], an established dataset with 51,839 images of German traffic signs annotated with sign type labels as ground truth (i.e., the intended outcome). Because the GTSRB data does not provide any further information that can be used to define semantic factors with a potential influence on uncertainty, e.g., location, time of day, or weather conditions, we had to augment the available data with a selection of realistic quality deficits and annotate them respectively.

To accomplish this, we applied the data augmentation framework proposed by Jöckel and Kläs [21], which allows augmenting an image with a selection of photorealistic quality deficits based on a given *situation setting*. About 2.7 million realistic situation settings were generated based on historical weather data from Deutscher Wetterdienst [22] and street locations within the TAS from OpenStreetMap [23].

In our study, we considered 9 types of *quality deficits* that could affect an image [24]: rain, darkness, haze, natural backlight, artificial backlight, dirt on the traffic sign, dirt on the sensor lens, steamed-up sensor lens, and motion blur. The intensity of each deficit was normalized to a scale between 0 (no effect) and 1 (maximum effect).

Each image of the original training dataset of GTSRB with 39,209 samples, which we refer to as the **basic** *training dataset*, was augmented for each quality deficit with low, medium, and high intensity by sampling from all appropriate situation settings and applying respective augmentations. This means each original image was augmented $9 \times 3 = 27$ times, provided sufficient situation settings where available. Besides the resulting 944,261 augmented and annotated images, the **augmented** *training dataset* additionally included the original 39,209 images of the basic training dataset.

The 12,630 images available in the GTSRB test dataset were randomly split into two equally sized, disjoint partitions for calibration and evaluation. To keep the original ratio between training and test data consistent, we augmented each original image 28 times based on settings randomly sampled from the 2.7 million realistic situation settings available. This resulted in 176,820 augmented samples each for the *calibration dataset* and the *evaluation dataset*. As the settings were generated based on the emulated target application scope, we assume that the random samples in this ***TAS-based*** *test dataset* has a distribution representative for the target application scope.

3.3 Study Execution

This section provides details on the four study execution steps illustrated in Fig. 2.

(1) Training of DDMs. To investigate our research questions, we considered two types of architectures for the DDMs: a single state-of-the-art convolution neural network (CNN) architecture with a softmax output layer, and a deep ensemble architecture including multiple CNNs running in parallel to represent what can be considered the current gold standard for in-model uncertainty estimations.

For the ***single CNN***, we chose a model architecture roughly based on the model that currently performs best in the GTSRB [1] in a variant without spatial transformers and using batch normalization in combination with spatial dropout instead of local contrast normalization after each convolution layer. These modifications were mainly motivated by reducing the computation resources that are required to build and evaluate deep ensembles based on several of these CNNs.

For the ***deep ensemble***, we chose an architecture that combines multiple CNNs of the same architecture and with the same hyper-parameter settings as the single CNN but with different weight initializations during model training. Following the conclusions of Henne et al. [19], who examined the effect of the number of ensemble members on the quality of uncertainty estimates, we decided to use seven ensemble members, which were identified as an adequate number in their study.

Motivated by our research questions, each DDM was trained either with the basic or the augmented training dataset. For the basic training dataset, prediction accuracy stabilized on holdout validation data for the trained CNNs around 0.995 after 30 epochs. Due to the much larger number of samples in our augmented training dataset, accuracy stabilized there after only 10 epochs at around 0.891.

(2) Calibration of DDMs. In uncertainty estimation research, calibration is seen as an important post-processing technique for in-model approaches to make their uncertainty estimates more reliable. In our study, we decided to apply the scikit-learn implementations of Isotonic Regression and Platt's logistic model, which are both model-agnostic, well-established calibration approaches, using the calibration dataset we had prepared for this purpose. When we subsequently talk about ***calibrated*** DDMs, we report the result based on Isotonic Regression, since Isotonic Regression consistently outperformed Platt's logistic model on our datasets. DDMs using a deep ensemble architecture are calibrated by calibration of the final outputs after combining the preference values of all ensemble members as proposed by [23].

(3a) In-model uncertainty estimation. In cases where we used in-model uncertainty estimates, the (calibrated) CNN or deep ensemble does not only report the predicted traffic sign type but also the uncertainty. In these cases, the uncertainty estimate is one minus the ultimately calibrated (and aggregated) preference value(s) as calculated by the softmax-layer(s) of the CNN(s).

(3b) Outside-model uncertainty estimation. Motivated by our research questions, we built UWs with three different sets of quality factors as input.

The default variant includes only *semantic* quality factors, meaning we consider as input the bounding box size around the detected traffic sign (assuming that images with fewer pixels make the task more difficult), the category of the predicted traffic sign (assuming that traffic signs of certain categories are more difficult to distinguish), and all nine types of augmented deficits (including rain, darkness, etc.).

The *combined* variant uses as an additional quality factor the 'in-model' uncertainty estimates computed by the encapsulated black-box DDM. Depending on the DDM, these uncertainty estimates can be raw softmax values or can be calibrated.

The third variant, the *computed* one, only considers quality factors as input that can be derived directly – without any additional information source–from the DDM input or output. In our setting, this comprises the size of the bounding box around the detected traffic sign, the DDM predicted traffic sign category and uncertainty.

The quality impact models of all UWs were trained as a decision tree built with CART algorithm optimized based on entropy with no pruning while training. After training, the quality impact models were calibrated on the calibration dataset considering a confidence level of 0.9999 and pruning all leaves containing less than 200 data samples in the case of UWs considering only semantic factors, and 700 otherwise (based on the results of a grid search).

Since our investigation focused on uncertainty related to input quality, which can be considered as the key strength of in-model uncertainty estimation approaches, a scope compliance model was not included and data points are considered inside TAS.

(4) Evaluation of uncertainty estimates. All uncertainty estimation approaches were evaluated on the TAS-based test dataset. To measure the uncertainty estimation performance, we computed the Brier score (*bs*), which measures the mean squared difference between the predicted probability of an outcome and the actual outcome [8]. The Brier score can be decomposed into variance (*var*), resolution (*res*), and unreliability (*unr*) [9] with $bs = var - res + unr$. A high *variance* corresponds to a high error rate of the DDM, i.e., more overall uncertainty. *Resolution* describes how much the case-specific uncertainty estimates differ from the overall uncertainty. As the *res* is bounded by the *var* and higher *res* values are better, we report instead $var - res$ as *unspecificity*. Finally, *unreliability* measures how well the estimated uncertainty is calibrated to the observed error rate of the DDM, i.e. smaller unreliability means better calibration. Additionally, we report as a metric of *overconfidence* the part of the unreliability attributed to uncertainty estimates that underestimate the observed error rate, which is the more serious case in a safety-critical setting.

4 Study Results and Discussion

We organized this section along the identified research questions RQ1 to RQ3, for which we will first present and then discuss the obtained evaluation results.

4.1 RQ1: Comparing UW Performance with In-Model Approaches

This section addresses the question of how uncertainty estimation performance between UWs and DDMs differs when they were trained and calibrated on appropriate data with good coverage of potential quality deficits. Accordingly, Table 2 presents the evaluation metrics for both DDM architectures – a *single* CNN and deep *ensemble* of CNNs – in comparison to the performance of UWs encapsulating these DDMs and relying in their estimates instead on *semantic* factors.

As Table 2 shows, the UWs based on semantic factors performed worse than the corresponding in-model approaches if we consider the Brier score as a global measure of performance. The main reason can be seen in their higher unspecificity, which is not completely compensated by their improved reliability. Please note that unreliability for the UWs was calculated considering a confidence level of .9999, which results in an intended unreliability increase as overconfident estimates are penalized. Had we not demanded this confidence level, the unreliability of the UWs would be two magnitudes lower (single = 0.00043, ensemble = 0.00039). Overconfidence was strongly decreased by UWs compared to in-model approaches.

Interpretation: In settings where appropriate training data is available, using UWs seems to be a trade-off between (a) reducing the resolution of uncertainty estimates and (b) obtaining higher interpretability, low overconfidence based on defined confidence levels, and separation of concerns through an outside-model approach.

Table 2. Study results on the performance of in-model approaches and UWs.

Research question	DDM archit	U. Estimation approach	Brier score	Variance	Unspecificity	Unreliability	Over-confidence
Baseline	Single	In-model	.09048	.19553	.00033	.09016	4.5e-02
RQ1		UW/*semantic*	.14931	.19553	.14630	.00301	2.4e-07
RQ2		UW/*combined*	.09065	.19553	.09018	.00048	8.9e-08
RQ3.1		UW/*computed*	.09076	.19553	.09048	.00028	1.6e-08
Baseline	Ensemble	In-model	.08584	.18696	.00034	.08550	3.8e-02
RQ1		UW/semantic	.14501	.18696	.14236	.00264	9.6e-07
RQ2		UW/*combined*	.08585	.18696	.08549	.00037	1.0e-06
RQ3.1		UW/*computed*	.08594	.18696	.08563	.00031	1.0e-06

Figure 3 provides an impression of the human interpretability of UW due to the use of semantic factors and their decision tree structure. For instance, the decision paths can be checked for plausibility (e.g., whether it is reasonable to assume that traffic signs are harder to classify if they are covered by dirt).

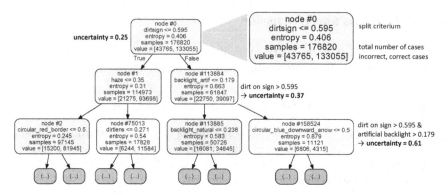

Fig. 3. Calibrated decision tree as part of an UW considering semantic factors.

4.2 RQ2: Synergies Between In-Model and Outside-model Approaches

This section addresses the question whether the uncertainty estimation performance of UWs can be improved if the in-model uncertainty estimates of the DDM are also considered as a quality factor, in addition to semantic factors such as precipitation.

The results in Table 2 show that if semantic factors are *combined* with in-model uncertainty estimates as an additional quality factor, the Brier score of the UWs almost reaches the level of the in-model approaches. The combination of in-model uncertainty estimates to the semantic factors not only strongly decreases the unspecificity but also further improves the reliability of the UW estimates.

Interpretation: Considering in-model uncertainty estimates as a factor can improve the uncertainty estimation performance of a UW up to the level of current gold standard in-model predictions. However, this performance improvement comes at the cost of decreased interpretability because a non-semantic "black-box" factor is introduced. Yet, other advantages are preserved, including confidence and separation of concerns.

4.3 RQ3: Performance Under Common, Less Than Optimal Conditions

This section investigates questions regarding the performance of UWs when applied under less than optimal conditions as they commonly occur in real-world settings.

RQ3.1. The results in Table 2 show that the uncertainty estimation performance if no semantic, but only *computed* factors are available in the UW reaches nearly the performance of considering the *combined* factors.

Interpretation: From a performance perspective, UWs with only computed quality factors work surprisingly well. However, we have to remember that neglecting semantic factors reduces interpretability, which is an important advantage of UWs. Thus, we conclude that UWs can also be applied when no semantic factors are available. However, developers should do as much as reasonably practical to assure interpretability, which includes considering as many semantic factors as possible.

RQ3.2. Table 3 shows how estimation performance is affected when the uncertainty estimates of the DDM, which were also used as a quality factor in the combined and computed UW, were not calibrated.

In the case where uncertainty was estimated by the *single* CNN, the Brier score became worse if the DDM was not calibrated. However, if the uncertainty estimates of the uncalibrated DDM were used instead as a quality factor of an UW, we observed no comparable negative impact on the performance of the encapsulating UW. The Brier score of the encapsulating UW (with *combined* and *computed* factors) was even better than the score of the encapsulated uncalibrated as well as calibrated single CNN. The last statement does also hold for the case that the DDM is the considered *deep ensemble*, which appears to have performed quite well even if not calibrated, but to a smaller magnitude.

Interpretation: Encapsulating an uncalibrated DDM within an UW seems to influence the components of the Brier score in a similar way as the calibration of the DDM. When DDMs are used encapsulated by an UW, the calibration of the DDM may thus not be required since its calibration has minor relevance for the performance of the UW.

RQ3.3. Table 4 summarizes the study results addressing the question of how the uncertainty estimation performance of (un-)calibrated DDMs and encapsulating UWs was affected when the DDMs were trained on data that did not sufficiently cover relevant quality deficits.

If the training dataset insufficiently covers relevant quality deficits, as the *basic* training dataset did in our setting, the uncertainty estimation performance as measured by the Brier score gets worse. This applies in particular when the DDM is additionally not calibrated on representative data. The worse Brier score results we obtained were caused by higher variance, higher unspecificity (except calibrated in-model approaches) and higher unreliably. If the in-model uncertainty estimates are calibrated, using an ensemble instead of a single CNN, does not improve the estimation results. The observed performance decrease was less noticeable for UWs compared to in-model uncertainty estimation approaches.

Interpretation: The training dataset is an important ingredient for obtaining reliable uncertainty estimates, especially for in-model approaches. In opposite to the case of uncalibrated DDMs, for which using an ensemble may compensate a missing calibration, in the case of calibrated DDMs that are trained on data insufficiently covering relevant quality deficits using an ensemble seems not to be an effective countermeasure. The observed negative effect of training data can however be partially mitigated by encapsulating the DDM within an UW or at least calibrating it.

RQ3.4. The results in Tables 2, 3 and 4 also compare the uncertainty estimation performance obtained when a single CNN or a deep ensemble of CNNs is used. The numbers show an improved Brier score in almost all cases when a deep ensemble was considered. The magnitude of improvement was especially high if the DDM was not calibrated. If the DDM was encapsulated, the performance gap between single and deep ensembles was smaller.

Table 3. Study results on using uncalibrated instead of calibrated DDMs.

DDM archit	DDM calibration	U. Estimation approach	Brier score	Variance	Unspecificity	Unreliability	Over-confidence
Single	Isotonic	In-model	.09048	.19553	.00033	.09016	4.5e-02
	-	In-model	.09481	.19946	.00155	.09326	6.1e-02
	Isotonic	UW/semantic	.14931	.19553	.14630	.00301	2.4e-07
	-	UW/semantic	.15071	.19946	.14755	.00316	1.6e-07
	Isotonic	UW/combined	.09065	.19553	.09018	.00048	8.9e-08
	-	UW/combined	.09013	.19946	.08967	.00046	0.0
	Isotonic	UW/computed	.09076	.19553	.09048	.00028	1.6e-08
	-	UW/computed	.09020	.19946	.08987	.00033	0.0
Ensemble	Isotonic	In-model	.08584	.18696	.00034	.08550	3.8e-02
	-	In-model	.08547	.18941	.00050	.08497	3.6e-02
	Isotonic	UW/semantic	.14501	.18696	.14236	.00264	9.6e-07
	-	UW/semantic	.14492	.18941	.14202	.00289	2.8e-07
	Isotonic	UW/combined	.08585	.18696	.08549	.00037	1.0e-06
	-	UW/combined	.08536	.18941	.08498	.00038	0.0
	Isotonic	UW/computed	.08594	.18696	.08563	.00031	1.0e-06
	-	UW/computed	.08540	.18941	.08503	.00038	0.0

Table 4. Results for DDMs trained on a basic dataset insufficiently covering quality deficits.

DDM archit	Training data	U. Estimation approach	Brier score	Variance	Unspecificity	Unreliability	Overconf
Single	Augmented	In-model/*cal*	.09048	.19553	.00033	.09016	4.5e-02
	basic	In-model/*cal*	.12116	.24025	.00001	.12116	4.3e-02
	Augmented	In-model/*uncal*	.09481	.19946	.00155	.09326	6.1e-02
	basic	In-model/*uncal*	.45116	.23491	.03360	.41756	4.1e-01
	Augmented	UW/*semantic*	.14931	.19553	.14630	.00301	2.4e-07
	basic	UW/*semantic*	.16438	.24025	.16127	.00311	1.6e-07
	Augmented	UW/*combined*	.09065	.19553	.09018	.00048	8.9e-08
	basic	UW/*combined*	.11620	.24025	.11545	.00075	0.0
	Augmented	UW/*computed*	.09076	.19553	.09048	.00028	1.6e-08
	basic	UW/*computed*	.12116	.24025	.12041	.00075	0.0
Ensemble	Augmented	In-model/*cal*	.08584	.18696	.00034	.08550	3.8e-02
	basic	In-model/*cal*	.12152	.24885	.00000	.12152	3.8e-02
	Augmented	In-model/*uncal*	.08547	.18941	.00050	.08497	3.6e-02
	basic	In-model/*uncal*	.14389	.24668	.00100	.14290	1.0e-01
	Augmented	UW/*semantic*	.14501	.18696	.14236	.00264	9.6e-07
	basic	UW/*semantic*	.17356	.24885	.17058	.00298	3.0e-06
	Augmented	UW/*combined*	.08585	.18696	.08549	.00037	1.0e-06
	basic	UW/*combined*	.11498	.24885	.11443	.00055	7.2e-08
	Augmented	UW/*computed*	.08594	.18696	.08563	.00031	1.0e-06
	basic	UW/*computed*	.11887	.24885	.11826	.00061	0.0

Interpretation: The better performance of deep ensembles in comparison to single CNNs comes at the cost of higher resource consumption during development and operation, which in our setting was seven times higher. The degree to which a deep ensemble exceeds the performance of a single model, and hence its cost-benefit ratio, has to be assessed on the specific application setting.

5   Conclusion

We compared UWs as an outside-model approach with existing in-model uncertainty estimation approaches. Our study focused on uncertainty estimation performance and related overconfidence considering the example task of traffic sign recognition.

Summarizing our conclusions on the posed research questions, we offer the following preliminary advices. (A1) If general uncertainty estimation performance is the only criterion, we recommend using a state-of-the-art *in-model* approach such as deep ensembles. (A2) If, additionally, computational resources are a limiting factor, a simpler DDM can be an alternative if its uncertainty estimates are calibrated appropriately. (A3) If at least one of the following criteria is relevant in our setting, the use of an UW, as an *outside-model* approach, should be considered:

- separating the concerns of providing good outcome and uncertainty estimates
- assuring interpretability, e.g., to check the plausibility of uncertainty estimates
- providing statistical guaranties based on a given confidence level
- facing a DDM trained on data insufficiently covering relevant quality deficits
- scope compliance is not guaranteed (e.g., causing out-of-distribution issues)

If encapsulating a DDM within a UW, (A4) calibrating the DDM does not appear to be necessarily required, (A5) the kinds of factors that are used in a UW – semantic, computed, or both – is a trade-off decision that should consider the specific needs regarding interpretability, available data, and estimation performance. In a safety-driven setting, we would recommend using semantic factors to the extent that is reasonably practical to keep the uncertainty estimates as transparent as possible.

In summary, we see good reasons to release AI models from the responsibility of providing dependable uncertainty estimates, and, considering the presented results, an uncertainty wrapper can be an option for realizing the required separation of concerns.

For the future, we plan to conduct additional studies to investigate the usefulness of the UW pattern in different settings and integrate it in a structured safety argument.

Acknowledgments. Parts of this work have been funded by the Observatory for Artificial Intelligence in Work and Society (KIO) of the Denkfabrik Digitale Arbeitsgesellschaft in the project "KI Testing & Auditing".

References

1. Arcos-García, A., Alvarez-Garcia, J., Soria Morillo, L.: Deep neural network for traffic sign recognition systems: an analysis of spatial transformers and stochastic optimisation methods. Neural Netw. **99**, 158–165 (2018)
2. Garcia-Garcia, A., Orts, S., Oprea, S., Villena Martinez, V., Rodríguez, J.: A review on deep learning techniques applied to semantic segmentation. arXiv:170406857 (2017)
3. Kläs, M.: Towards identifying and managing sources of uncertainty in AI and machine learning models - an overview. arXiv:1811.11669 (2018)
4. Dijkstra, E.W.: On the role of scientific thought. In: Selected writings on Computing: A Personal Perspective, pp. 60–66. Springer, New York, USA (1982). https://doi.org/10.1007/978-1-4612-5695-3_12
5. Kläs, M., L. Sembach, L.: Uncertainty wrappers for data-driven models–increase the transparency of AI/ML-based models through enrichment with dependable situation-aware uncertainty estimates. In: WAISE (2019)
6. Kläs, M., Vollmer, A.M.: Uncertainty in machine learning applications – a practice-driven classification of uncertainty. In: WAISE (2018)
7. Kläs, M., Jöckel, L.: A framework for building uncertainty wrappers for AI/ML-based data-driven components. In: WAISE (2020)
8. Brier, G.W.: Verification of forecasts expressed in terms of probability. Mon. Weather Rev. **78**(1), 1–3 (1950)
9. Murphy, A.H.: A new vector partition of the probability score. J. Appl. Meteorol. **12**(4), 595–600 (1973)

10. Kläs, M., Adler, R., Sorokos, I., Joeckel, L., Reich, J.: Handling uncertainties of data-driven models in compliance with safety constraints for autonomous behavior. In: European Dependable Computing Conference (EDCC), (2021, accepted for publication)
11. Der Kiureghian, A., Ditlevsen, O.: Aleatory or epistemic? Does it matter? Struct. Saf. **31**(2), 105–112 (2009)
12. Bandyszak, T., Jöckel, L., Kläs, M., Törsleff, S., Weyer, T., Wirtz, B. Handling uncertainty in collaborative embedded systems engineering. In: Böhm, W., Broy, M., Klein, C., Pohl, K., Rumpe, B., Schröck, S. (eds.) Model-Based Engineering of Collaborative Embedded Systems, pp. 147–170. Springer, Cham (2021). https://doi.org/10.1007/978-3-030-62136-0_7
13. Guo, C., Pleiss, G., Sun, Y., Weinberger, K.: On calibration of modern neural networks. In ICML (2017)
14. Pimentel, M., Clifton, D., Clifton, L., Tarassenko, L.: A review of novelty detection. Sig. Process **99**, 215–249 (2014)
15. Aslansefat, K., Sorokos, I., Whiting, D., Tavakoli Kolagari, R., Papadopoulos, Y.: SafeML: safety monitoring of machine learning classifiers through statistical difference measures. In: IMBSA (2020)
16. Arnez, F., Espinoza, H., Radermacher, A., Terrier, F.: A comparison of uncertainty estimation approaches in deep learning components for autonomous vehicle applications. In: AISafety (2020)
17. Lakshminarayanan, B., Pritzel, A., Blundell, C.: Simple and scalable predictive uncertainty estimation using deep ensembles. In: NIPS (2017)
18. Gustafsson, F., Danelljan, M., Schön, T.: Evaluating scalable Bayesian deep learning methods for robust computer vision. In: CVPR (2020)
19. Henne, M., Schwaiger, A., Roscher, K., Weiss, G.: Benchmarking uncertainty estimation methods for deep learning with safety-related metrics. In: SafeAI (2020)
20. German Traffic Sign Benchmarks (2021). http://benchmark.ini.rub.de/?section=gtsrb
21. Jöckel, L., Kläs, M.: Increasing trust in data-driven model validation. In: SafeComp (2019)
22. Climate Data Center. https://cdc.dwd.de/portal/. Accessed 13 Nov 2020
23. OpenStreetMap. https://www.openstreetmap.de/. Accessed 13 Nov 2020
24. Jöckel, L., Kläs, M., Martínez-Fernández, S.: Safe traffic sign recognition through data augmentation for autonomous vehicles software. In: QRS (2019)
25. Rahaman, R., Thiery, A.: Uncertainty quantification and deep ensembles. arXiv:2007.08792 (2020)

Towards Certification of a Reduced Footprint ACAS-Xu System: A Hybrid ML-Based Solution

Mathieu Damour[1,2], Florence De Grancey[2,3], Christophe Gabreau[2,4],
Adrien Gauffriau[2,4], Jean-Brice Ginestet[5], Alexandre Hervieu[5],
Thomas Huraux[1,2], Claire Pagetti[6(✉)], Ludovic Ponsolle[2,7],
and Arthur Clavière[8]

[1] Scalian, Toulouse, France
[2] IRT Saint Exupéry, Toulouse, France
[3] THALES, Paris, France
[4] Airbus, Toulouse, France
[5] DGA, Balma, France
[6] ONERA, Toulouse, France
claire.pagetti@onera.fr
[7] Apsys, Blagnac, France
[8] Collins Aerospace, Toulouse, France

Abstract. Approximating while compressing lookup tables (LUT) with a set of neural networks (NN) is an emerging trend in safety critical systems, such as control/command or navigation systems. Recently, as an example, many research papers have focused on the ACAS Xu LUT compression. In this work, we explore how to make such a compression while preserving the system safety and offering adequate means of certification.

1 Introduction

Due to the intensive flights traffic, the risk of collision is increasing. During the last decade, a standardization group has defined a new competitive and effective anti-collision system named ACAS X (for *Next-Generation Airborne Collision Avoidance System*) [16]. The purpose is to keep any intruder outside of the desired envelope of the ownship.

1.1 ACAS Xu Overview

Among the family of ACAS X, we will focus on the ACAS Xu [10] dedicated to drone, Urban Air Mobility and Air Taxi with horizontal automatic resolution. The system is based on a set of lookup tables (LUT) that are used in real-time to resolve conflicts. Those LUT have been computed off-line and their size has been chosen in order to fulfil real-time (decisions must be taken every second) and safety level (there should not be any collision) requirements. The ownship computes six parameters (listed below) that enable to access the tables which

© Springer Nature Switzerland AG 2021
I. Habli et al. (Eds.): SAFECOMP 2021, LNCS 12852, pp. 34–48, 2021.
https://doi.org/10.1007/978-3-030-83903-1_3

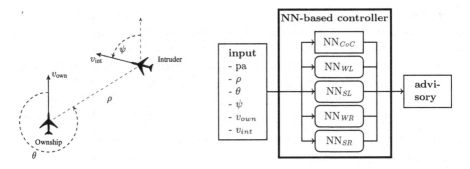

Fig. 1. ACAS Xu geometry [14]   **Fig. 2.** NN-based architecture [14]

give an estimation of the probability to have a collision for the each possible advisory and the chosen advisory is the one that minimizes this probability. The geometry of the system is given in Fig. 1, and the definition of the parameters stands as:

- ρ (ft): Distance from ownship to intruder
- θ (rad): Angle to intruder relative to ownship heading
- ψ (rad): Heading angle of intruder relative to ownship heading direction
- v_{own} (ft/s): Speed of ownship
- v_{int} (ft/s): Speed of intruder
- τ (s): Time until loss of vertical separation.

The 23 LUT provide the transitions costs between the previous advisory (pa) and the next advisory. There are five advisories: COC (Clear Of Conflict); SR (Strong Right); SL (Strong Left); WR (Weak Right) and WL (Weak Left). In particular, when the ownship is in the COC state, it can continue its mission. When the ownship is in one of the other states, it has to initiate a turn with a rate that depends of the computed state. In practice, a single table is composed of 2 sub-tables: the first contains definition of parameters values and the second contains the costs that are half-integer (16 bits). More information on the ACAS Xu system can be found in [19].

1.2 Purpose of the Work

Several universities have worked on replacing the LUT by neural networks (NN), the objective being to reduce the size of the embedded code and improve the anti-collision performance. The authors of [14] have replaced the LUT by 45 neural networks leading to an impressive reduction of the memory footprint (4 GB to 150 MB), see the Fig. 2 for horizontal advisory (i.e. when $\tau = 0$). In this work, we want to explore how to compress the LUT with neural networks while preserving the system safety and offering adequate means of certification. We only focus in the sequel on the horizontal resolution of conflict.

Certification Problem Statement. For any safety-critical system embedded in an aircraft, airframers (applicants) have to demonstrate to regulation authorities that their product is compliant with certification specifications. To this purpose, applicants use a set of standards that are recognized as acceptable means of compliance. Existing development assurance standards are not adapted to the data-driven paradigm of the ML technique, though such development may introduce errors that could jeopardize a safe operational use of the system (e.g. such probabilistic approach may introduce unforeseen errors). Currently a joint working group, the EUROCAE WG-114/SAE G-34 [9] (WG-114 for short in the sequel) is preparing the next standard to fill this gap.

Contributions. The use of NNs to approximate the LUT may lead to unexpected behaviors that should be mitigated to guarantee that the ML-based item will not alter the safety of the system. For this purpose, the ACAS Xu subsystem is designed as an hybrid controller: a non ML item is introduced to guarantee the safety in all the operational domain (safety net).

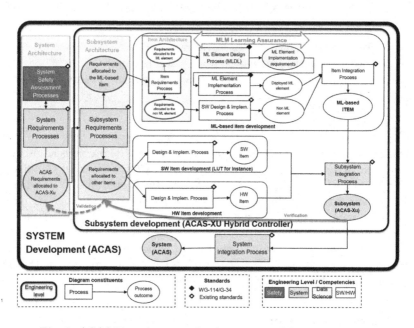

Fig. 3. ACAS Xu subsystem development workflow using ML

In addition, a new certification strategy is investigated to provide sufficient guaranties to authorities. The Fig. 3 instantiates the WG-114 development process workflow to the ACAS Xu use case. There are three levels of engineering: the *System and Subsystem Level* as proposed by ARP4754A [25] standard which provides guidance for the system development process and are complementary to the product requirements (WG 75-1 [10] for the ACAS Xu technical requirements); the *item Level* where ML activities are not covered by any known standard while the non-ML parts are supported by classical guidance for the implementation process of the software items (DO-178C [8]) and hardware items (DO-254 [23]).

To tackle the objectives of the system development guidance, three aspects were developed. The first concerns the learning assurance activity of the ML element (or MLM-Machine Learning Models), which aims at ensuring that the MLM requirements (covering functional, performance and robustness aspects) have been captured and correctly designed. The elements supporting this activity are detailed in Sect. 2. The second aspect (dashed green arrow) targets the item validation activity in order to check that captured requirements fit the system needs. The third aspect (solid green arrow) targets the subsystem verification to check that the ACAS-Xu Hybrid controller safely performs its intended use. Activities and relationships covering the two first aspects are part of a constructed argumentation (see Sect. 3).

2 ACAS Xu Hybrid Architecture

We propose to replace the LUT with a hybrid architecture (shown in Fig. 4) composed of a neural networks based controller part (as proposed by [14]) together with a safety net to ensure a correct behavior. The idea, that will be detailed hereafter, is to 1) apply the best practices of learning assurance to well approximate the tables; 2) identify off-line the zones where the NN-based system differs from the LUT advisories and where it may jeopardize the safe behavior of the system; 3) compute on-line in which zone the system is, call the NN-based system if it behaves similarly to the LUT or switch to the safety net if not in order to always be safe (this step is performed by *check module*). The safety net consists of the extract of the LUT for these zones.

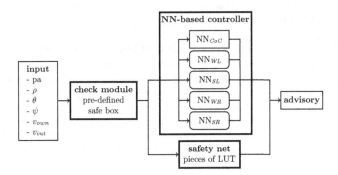

Fig. 4. Architecture of the neural network based ACAS Xu

2.1 Learning Process

The objective of the learning process is to build a model for advisory computation which performs a trade-off between reducing the memory footprint and preserving fidelity to LUT. Ideally, this model should reconstruct the original (but unknown) cost function. NNs are quite good universal approximators as long as the cost function has similar bounded derivative on the whole domain which is unfortunately not the case here. The cost function shows two local offsets where derivative reaches high value: one (Fig. 5) is observed for each cost function when the range is below 5000 ft, where the cost functions switches quickly from 0 to 15000, and the second (Fig. 6) concerns the case of COC advisory cost value when previous advisory is not COC, where a local offset of +4000 is added. This issue could be resolved by either augmenting the size of the NNs (not explored here) or finding the most suitable NN architecture.

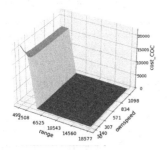
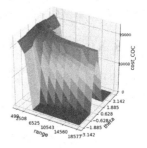

Fig. 5. Cost function CoC → CoC   **Fig. 6.** Cost function WL → CoC

We explored several architectures (with ReLu activation only) with the same learning process where 1) input and output data are normalized between -1 and 1 (as suggested in [17]), 2) batches size is set to 8912, 3) Adam optimizer [5] was used, 3) initial learning rate is set to 0.002:

1. regression (cost inference) versus classification (decision inference);
2. regular hidden layer size versus decreasing layer size. We have tested *regular50* with 8 layers and 5-50-50-50-50-50-50-5 neurons per layer; *decreasing128* as 5-128-64-32-16-5 and *decreasing256* as 5-256-128-64-32-16-5.

The criteria for evaluation, that we called *accuracy*, is the agreement rate between advisory computed by the NN and the ground truth LUT value. Both training and evaluation were performed in the whole data-set which is not classical but the point is to be as close as possible to the LUT. Thus, for once overfitting is encouraged to improve accuracy.

We completed the evaluation with accuracy measurement on range restricted subsets which represent the most critical situations. At long-range, as there is no risk of collision and the advisory is most of the CoC (95% of advisory for ranges above 10000 ft). For ranges below 5000 ft, the advisory is spread with 15% CoC, 38% SL and 36% SR. We set empirically two subsets: a mid-range subset with range below 20000 ft, and a short range subset with ranges below 500ft. The table above shows the measured accuracy.

Training type	Network shape	Accuracy	Accuracy mid range	Accuracy short range	Nparam
Regression	[14]	93,22	82,424	68,25	13305
Regression	regular50	95,42	87,44	71,52	15855
Regression	decreasing128	95,81	89,18	75,62	28229
Regression	decreasing256	96,33	90,82	79,74	111173
Classification	decreasing256	76,06	76,08	86,43	111173

Overall, as expected we observe that accuracy decreases when the range is reduced. Furthermore, at very short ranges, we observe in LUT that the cost difference between decision is very low, then it is more difficult for neural network to infer the exact cost. We observe that the regression approach is more effective than classification. This result could be explained by the fact that the decision boundary have "square" shape, not suitable for shallow neural networks. We also observe that the decreasing architecture performs better than the regular architecture. We can suppose that this decreasing architecture favours representation of more complex functions in the first layers. Such complex representation is more suitable to represent offset effects. After this study we have selected regular50 and decreasing128 since they reaches high accuracy with limited memory (parameters) footprint.

2.2   Design of the Hybrid Architecture

The objective of the safety net is to take over when the NN does not take similar advisory as the LUT in the same situation. To determine the zones where the NN differ from the LUT, we use formal verification techniques. More precisely, we decompose the space as a set of p-dimensional boxes (short as p-box).

Definition 1 (p-box). *Let $p \in \mathbb{N}$, a p-dimensional box $[b]^p$ is a set of $\mathbb{R}^p$ defined as the cartesian product of p intervals:*

$$[b]^p = \underset{1 \leq i \leq p}{\times} [l_i, u_i]$$

wherein $l_i \in \mathbb{R}$ (resp $u_i \in \mathbb{R}$) is the lower bound (resp the upper bound) of the ith interval composing the box $[b]^p$.

The boundaries of the p-boxes come from the parameters values of the LUT. In practice, the input state space is split in $36,784.10^6$ 5-boxes. For each box, we compute the possible decisions obtained either from the LUT and the NN, and we check that they are *similar*, which is formalized in the property below.

Property 1 (NN-based architecture compliant with specification). We define by decisions $f(l) \subseteq \{CoC, WL, SL, WR, SR\}$ the set of reachable advisories by f from any point of l where $f \in \{NN, LUT\}$ and l is a p-box. We consider that a NN behaves similarly to the LUT on an p-box l if

$$decisions\ NN(l) \subseteq decisions\ LUT(l)$$

To determine which p-box satisfies the Property 1, we use the verification tools DEEPPOLY [28], RELUPLEX [14] and PLANET [7]. First the Property 1 is checked on the p-boxes with DEEPPOLY. As DEEPPOLY computes an over-approximation, thus either DEEPPOLY provides a positive answer (i.e. property holds) or an unknown answer. Then RELUPLEX/PLANET is called on the remaining boxes (those for which DEEPPOLY provided unknown).

In this work, the safety net is designed with the ownship and intruder having a constant speed of 438 ft/s and 414 ft/s respectively. This corresponds most of the time to the worst case situation, i.e. if an aircraft flights slower, decisions are similar. This improves the compression of the hybrid architecture and speeds up the safety net design. In effect, we would need to design the safety so that it covers all situations. Because of our hypotheses ($\tau = 0$ for horizontal resolution and constant speed), the space is split in 304 000 3-boxes and thus one 3-box is defined by (ρ, θ, ψ). The results are given in the table below.

Method	DEEPPOLY		RELUPLEX/PLANET time	Number of failed boxes
	Time	Success		
Regression [14]	14 min	77.2%	63 h	24057
regular50	15 min	78.7%	48 h	6912
decreasing128	16 min	78.2%	56 h	1664

We plot (in Fig. 7) for each 3-box (ρ, θ, ψ) the lower bound l_i with on the left a color indicating the solver used (RELUPLEX/PLANET or DEEPPOLY) and on the right a color indicating the advisory. More precisely, right plot of the figure shows advisory for an intruder aircraft located at each point on the plot, whose coordinates indicate the slant range (ρ) and angle to intruder (θ) and with the own-ship located at the center of the plot. Since 87% of LUT(l) decisions are unique, it entails that the NNs take exactly the same decision as the LUT most of the time.

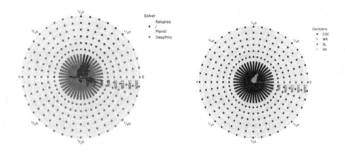

Fig. 7. Solver used for proving properties over 3-box - Polar coordinates (ρ, θ)

We can see that DEEPPOLY is able to quickly prove properties in area where the cost functions are very different, whereas it does not reach a proof in areas where cost functions are very close. Difficult verification needing RELU-PLEX/PLANET are the areas where ACAS Xu system gives avoidance orders.

2.3 Why a New Hybrid Architecture

Current implementation of ACAS Xu should embed 4 Gbytes LUT and executes 1 Hz. Such implementation with avionic constraints is rather challenging, see for instance [21]. In particular, there is not much such large memory available on the market that is compatible with avionics constraints. Compression is therefore a strategic approach but not at the cost of reduced safety. This is the reason why we approximate them as NN together with a safety net. Using an approved fallback to mitigate safety risks is regularly used in the avionic system architectures. This is also a strong recommendation of the AVSI report [2] to bound the behavior of ML algorithms and prevent any unintended behaviour that may challenge the system safety.

In the table below, we have computed the size needed by the neural networks as well as the one for managing the switch and the safety net.

Network shape	Nb of parameters in NN	NNs size (MB)	Failed boxes (kB)	Full memory footprint (MB)
[14]	598,725	102.6	564.0	103.2
regular50	713,475	122.4	162.0	122.7
decreasing128	1,270,305	217.8	39.0	217.8

For the safety net and *check module*, we need to store p-boxes and exactly the same since *check module* identifies when to switch in the safety net. Each unsafe box will be stored in the memory using the lower and upper points. Using float32, the size needed by a 3-boxes is 24 bytes. This leads to a size of 974 kB for [14], 4.1 kB for regular50 and 4.0 kB for decreasing128. Footprint of unsafe boxes is one order of magnitude below the networks footprint and is decreasing with the size of the network. The compression of NN could also be improved using pruning and quantization techniques [11].

3 Certification Methodology

Assurance cases (AC) are gaining more and more consideration as valuable methodologies for development and certification. John Rushby [24] defines them as: *Assurance cases are a method for providing assurance for a system by giving an argument to justify a claim about the system, based on evidence about its design, development, and tested behavior.*

3.1 Notations

The idea is to detail the argumentation leading to a certain *conclusion* or *claim*. In the context of certification, a claim is an objective to be fulfilled by the

applicant. In practice, the demonstration is based on the elicitation of requirements that correspond to the justifications that the objective is achieved. There
exist several notations, either textual or graphical, to support the design of an
assurance case, such as GSN (Goal Structuring Notation) [15]. All of them are
relying on the Toulmin work [29].

Among the existing notations, we will use subsequently in the paper a graphical adaptation of Toulmin notation proposed by the RESSAC [22] project.
RESSAC was a European project that coordinated European industry efforts
to contribute to the FAA initiative called "Overarching Properties", which pro-
motes an alternative certification approach to ease the introduction of next generation systems. This notation relies on (see Fig. 8): the Claim C is either the
upper conclusion or intermediate conclusions (or sub-claims), the Evidence E
is a leaf that consists of a V&V documentation that supports some claim, the
Reasoning R explicitly describes the argument and the Backing B supports the
reasoning. The Backing is a kind of endorsement of the reasoning, a guarantee
that the reasoning is reliable. The defeater D allows for expressing that in some
circumstances the conclusion may not be true. Such a notation is very helpful
as it offers simplicity and the possibility to challenge the reasoning steps. The
context contains additional information needed to provide definitions or descriptions of terms constraining the applicability of the assurance case to a particular
environment or set of conditions.

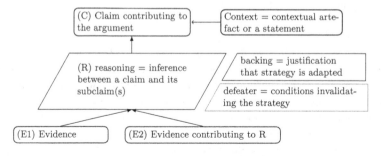

Fig. 8. Graphical RESSAC notation

3.2   Assurance Case for the Hybrid Controller

The objective of the certification approach is two-fold: first demonstrate the
completeness and the correctness of the ML-based item implementation with
respect to the system and safety requirements; and second reinforce the confidence that the ML-based item has been developed in a sufficiently disciplined
manner to limit the likelihood of development errors that could impact system
safety. This approach covers the "Learning Assurance" and "AI safety risk mitigation" building blocks, pillars of the trustworthiness concept introduced by the
EASA AI Roadmap 1.0 [6].

The overall assurance case for the hybrid architecture is quite large as it covers the full ACAS system development and contains 120 elements (claims, reasoning, context, backing and evidences) addressing the objectives of the ARP4754A [25]. Due to space limit, we cannot detail everything and we chose to focus on some objectives.

ACAS-Xu Subsystem Requirement Capture. The Fig. 9 shows the reasoning to demonstrate that the ACAS Xu specification process meets the ARP4754A [25] guidance concerning the definition of system requirements and interfaces. The argument is based on 2 sub-claims: the capture of the functional and performance requirements. The performance requirements are not further detailed.

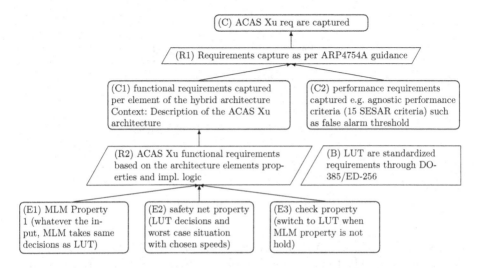

Fig. 9. Assurance case - ML subsystem requirements

The functional requirements must be refined for each item of the hybrid architecture (NNs and safety net). The reasoning is that the LUT decisions are the behaviour reference of the controller. Thus each item of the hybrid architecture should have equivalent properties and the switch logic should be appropriate. Specifically, property 1 is defined to guarantee the correct operation of the MLM. **ACAS-Xu Item Verification (for the ML Element Robustness Part).** As per [3], one of the main premises of the robustness demonstration is "*real-world situations to which the subsystem is not robust should be identified and mitigated*" (refer to claim C in Fig. 10). All the situations where the MLM provides incorrect predictions (i.e. where Property 1 is not preserved by the MLM), are identified. The mitigation is realized by the architecture design (switch to the safety net which embeds the subset of LUT needed for mitigation). The preservation of the Property 1 is formally verified within the robustness analysis: the input space is divided into boxes defined by points of the LUT. When decisions associated to the top-points of a box are different from one another (frontiers of decisions), then the estimated prediction of each point of the box is considered as correct when identical to one of the box top points. Property 1 is verified using formal methods: when Property 1 does not hold, this means that the situation may be unsafe and that the hybrid controller should switch to the LUT computation to take the appropriate decision.

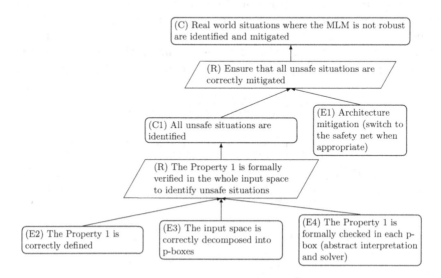

Fig. 10. Assurance case - ML item robustness

ACAS-Xu Subsystem Validation. Considering that LUT standardization has been recognized by the Authorities, one can think that the proper verification of the MLMs (to correctly approximate LUT predictions) would be sufficient to consider them as validated. Actually, the lack of transparency of the ML technique (no traceability capability, black box effect) may require a need for additional assurance that MLMs properties are correct and complete. For this purpose, specific tests have been developed in a simulation environment enabling the comparison between operational behaviors of ML-based design and real LUT design whatever the geometric situation. The Fig. 11 develops the argumentation and illustrates the use of defeater (D) to challenge the confidence that the use of standardized data may not be sufficient for demonstration of conformity.

4   Related Work

Proved ACAS Xu. Up to now, the ACAS Xu compression works only proposed to replace the LUT with a set of neural networks. Even if there have been several papers on formal verification of NN, none of them has tackled the certification itself. Most of the time, papers prove some local properties on the neural network which is not sufficient to cover certification expectation. Authors

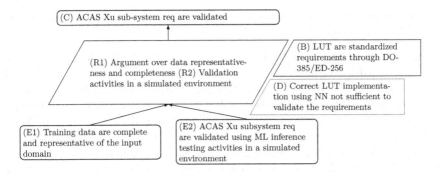

Fig. 11. Assurance case - ML subsystem validation

of [14] proposed to prove 10 avoidance meta-properties without any explicit link with avoidance standard [10]. We believe that these properties are not enough for enabling the certification of an ACAS-Xu system. Our approach is different, because we consider LUT as the requirements (part of the standard) and we formally guarantee that outputs of our system will be exactly the same as LUT.

Certification Methodology. There are several works offering assurance case to summarize confidence for ML components. [3] proposed a pattern to ensure the robustness of ML subsystems. We have completed and adapted this approach to integrate the specific properties (safe behaviour reference given by the LUT) and the safety net. [26] proposed a template to structure the safety argumentation part specific to DNNs. Their work is illustrated with an example use case based on pedestrian detection.

To the best of our knowledge, no assurance case approach has been proposed to tackle the respect of functional and performance objectives at system level (when ML sub-components are involved) for aircraft certification. The literature in the automotive is richer. In particular, [20] argued that assurance cases can be used for DNNs based systems. [30] goes further as it proposes GSN patterns to reason on the safety requirements of ML-based components and their integration within a system-level reasoning to show the compliance with ISO 26262. Thus, our work is complementary as we address the aeronautical sector and tackle the ARP4754A.

[1] proposed a novel concept of Dynamic Assurances Cases (DAC) that is applied to an aviation system that integrates ML-based perception function for autonomous taxiing. This concept is based on a framework of assurance methods /tools addressing safety concerns during development and extending this level of assurance to an in-flight operational use. They use both Assurance Case (with GSN notation) and architecture mitigation to develop assurance components for the DAC framework. Though we share the ARP4754A objectives and the assurance case methodology, our main objective is more to bridge the gaps of conformity of a ML-based system to the ARP4754A safety, functional and

operational objectives and guarantee an acceptable means of compliance with certification requirements.

Learning a Surrogate NN of a LUT. Neural networks are a new trend for approximating complex functions as a replacement of LUT, for example for control command systems [27]. Significant work has been performed in the framework of deep Q-learning, where the Q table is approximated by neural network. There have been some experiments in the context of Calibration Look-up table for the tuning of voltage-controlled circuits [18]. It was also explored for ACAS-Xu use case in [12,13]. In [13], authors compares table compression using the origami algorithm which exploits data's redundancies and symmetries, and a method and using a neural network with a regular architecture. In [12], further exploration of the neural network approach is performed, introducing several tricks to enhance performance.

5   Conclusion

We have designed a safe NN-based ACAS Xu architecture and shown with an assurance case that such a way of doing could be well argued for certification to the regulation authorities. The certification evidences will be completed with sub-system level analyses (with simulation and reachability analysis [4]).

In the future, we also plan to implement the hybrid architecture on an embedded board to complete the certification proof. We will also apply our methodology for other LUT-based safety critical systems.

Acknowledgments. This project received funding from the French "Investing for the Future – PIA3" program within the Artificial and Natural Intelligence Toulouse Institute (ANITI). The authors gratefully acknowledge the support of the DEEL project (https://www.deel.ai/).

References

1. Asaadi, E., et al.: Assured integration of machine learning-based autonomy on aviation platforms. In: 39th Digital Avionics Systems Conference (DASC 2020) (2020)
2. AVSI: Final Report AFE 87 - Machine Learning (2020)
3. Hawkins, R., Calinescu, R., Picardi, C., Paterson, C., Habli, I.: Argument patterns and processes for machine learning in safety-related systems. University of York, York, U.K. (2020)
4. Clavière, A., Asselin, E., Garion, C., Pagetti, C.: Safety verification of neural network controlled systems. In: 7th International Workshop on Safety and Security of Intelligent Vehicles (SSIV 2021) (2021)
5. Diederik, J.B., Kingma, P.: Adam: a method for stochastic optimization. In: 3rd International Conference for Learning Representations (2015)
6. EASA: Artificial Intelligence Roadmap: A human-centric approach to AI in aviation (2020)

7. Ehlers, R.: Formal verification of piece-wise linear feed-forward neural networks. CoRR, abs/1705.01320 (2017)

8. EUROCAE/RTCA: DO-178C/ED-12C - Software Considerations in Airborne Systems and Equipment Certification (2011)

9. EUROCAE WG-114/SAE Joint Group: Certification/approval of aeronautical systems based on AI (2021). On going standardization

10. EUROCAE WG 75.1 /RTCA SC-147: Minimum Operational Performance Standards For Airborne Collision Avoidance System Xu (ACAS Xu) (2020)

11. Han, S., Mao, H., Dally, W.J.: Deep compression: compressing deep neural networks with pruning, trained quantization and Huffman coding (2016)

12. Julian, K.D., Kochenderfer, M.J., Owen, M.P.: Deep neural network compression for aircraft collision avoidance systems. arXiv:1810.04240 (2018)

13. Julian, K.D., Lopezy, J., Brushy, J.S., Owenz, M.P., Kochenderfer, M.J.: Deep neural network compression for aircraft collision avoidance systems. In: 35th Digital Avionics Systems Conference (DASC) (2016)

14. Katz, G., Barrett, C.W., Dill, D.L., Julian, K., Kochenderfer, M.J.: Reluplex: An efficient SMT solver for verifying deep neural networks. CoRR, abs/1702.01135 (2017)

15. Kelly, T., Weaver, R.: The goal structuring notation /- a safety argument notation. In: Workshop on Assurance Cases (2004)

16. Kochenderfer, M., Holland, J., Chryssanthacopoulos, J.: Next generation airborne collision avoidance system. Lincoln Lab. J. **19**, 17–33 (2012)

17. LeCun, Y., Bottou, L., Orr, G., Müller, K.: Efficient backprop. In: Neural Networks: Tricks of the Trade, chap. 2, p. 546 (1998)

18. Leoni, A., Marinković, Z., Pantoli, L.: On the introduction of neural network-based optimization algorithm in an automated calibration system. In: 14th International Conference on Advanced Technologies, Systems and Services in Telecommunications (TELSIKS), pp. 323–326 (2019)

19. Manfredi, G., Jestin, Y.: An introduction to ACAS Xu and the challenges ahead. In: 35th Digital Avionics Systems Conference (DASC 2016), pp. 1–9 (2016)

20. Kaur, R., Ivanov, R., Cleaveland, M., Sokolsky, O., Lee, I.: Assurance case patterns for cyber-physical systems with deep neural networks. In: Casimiro, A., Ortmeier, F., Schoitsch, E., Bitsch, F., Ferreira, P. (eds.) SAFECOMP 2020. LNCS, vol. 12235, pp. 82–97. Springer, Cham (2020). https://doi.org/10.1007/978-3-030-55583-2_6

21. Ren, L., et al.: Integration and flight test of small UAS detect and avoid on a miniaturized avionics platform (2019)

22. RESSAC: Recommendations for the use of assurance cases for demonstrating and assessing overarching properties. Technical report, LIV-S026-D4-199 (2019)

23. RTCA, Inc.: DO-254 - Design Assurance Guidance For Airborne Electronic Hardware (2005)

24. Rushby, J.: The interpretation and evaluation of assurance cases. Technical report (2015). Technical Report SRI-CSL-15-01

25. E. SAE: Aerospace Recommended Practices ARP4754a/ed-79a- development of civil aircraft and systems (2010)

26. Schwalbe, G., et al.: Safety argumentation for deep neural network based perception in automotive applications. In: SAFECOMP 2020 Workshops (2020)

27. Seren, C., Ezerzere, P., Hardier, G.: Model-based techniques for virtual sensing of longitudinal flight parameters. Int. J. Appl. Math. Comput. Sci. **25**, 03 (2015)

28. Singh, G., Gehr, T., Püschel, M., Vechev, M.: An abstract domain for certifying neural networks. Proc. ACM Program. Lang. **3**(POPL) (2019)

29. Toulmin, S.E.: The Uses of Argument. Cambridge University Press, Cambridge (2003). Updated Edition, first published in 1958

30. Wozniak, E., Cârlan, C., Acar-Celik, E., Putzer, H.J.: a safety case pattern for systems with machine learning components. In: Casimiro, A., Ortmeier, F., Schoitsch, E., Bitsch, F., Ferreira, P. (eds.) SAFECOMP 2020. LNCS, vol. 12235, pp. 370–382. Springer, Cham (2020). https://doi.org/10.1007/978-3-030-55583-2_28

Security Engineering

IT Design for Resiliency Using Extreme Value Analysis

Szilárd Bozóki$^{(\boxtimes)}$ and András Pataricza

Department of Measurement and Information Systems, Budapest University of
Technology and Economics, Magyar Tudósok Krt. 2, Budapest 1117, Hungary
bozoki@mit.bme.hu, pataricza.andras@vik.bme.hu

Abstract. Safety-critical systems are designed to operate millions of
hours without losing or harming life. Millions of hours enable events
with small occurrence probability to materialise. Owing to this, rare
events have to be factored when designing for millions of safe operating
hours. In this paper, we apply a statistical paradigm named Extreme
Value analysis for the modeling of the rare events and probabilistic risk
assessment. Without loss of generality, our motivation is cyber-physical-
systems where the IT infrastructure is frequently shared between func-
tionally independent tasks and the run-time platform, such as Industry
4.0 based on 5G and edge cloud computing.

As a practical example, we present our method on a case study on
a typical micro-service-based edge computing setup by measuring and
analysing the container restart times in Kubernetes. The results can be
used to asses and compare resilience mechanism design alternatives.

Keywords: Design for resiliency · Extreme value analysis · Edge
computing · Statistical modeling · Dimensioning · Kubernetes

1 Introduction

Safety-critical systems appear in many forms and shapes, but they are all
designed to minimise and prevent harming life. In this paper, we focus on IT
systems where failure could lead to serious financial consequences.

Our textbook example is an edge cloud computing IT infrastructure. Edge
computing promises to reduce end-to-end-latency by leveraging low-latency 5G
radio networking and bringing computing power closer to the consumer via geo-
graphically distributed edge computing sites. This low-latency environment supports *soft/hard -real-time cyber-physical-systems* (CPS), such as Industry 4.0.

This paper partially relies on a previous joint project with Ericsson. Additionally, the
research reported in this paper and carried out at the BME has been supported by
the NRDI Fund based on the charter of bolster issued by the NRDI Office under the
auspices of the Ministry for Innovation and Technology and a funding from the EU
ECSEL JU under the H2020 Framework Programme, JU grant nr. 826452 (Arrowhead
Tools project) and from the partners' national funding authorities.

ⓒ Springer Nature Switzerland AG 2021
I. Habli et al. (Eds.): SAFECOMP 2021, LNCS 12852, pp. 51–66, 2021.
https://doi.org/10.1007/978-3-030-83903-1_4

Dimensioning (resource planning) will be critical to edge computing sites because their remote nature increases operational and resource costs. Additionally, compared to regular web services hosted on regular clouds, the applications edge computing intends to serve (e.g. Industry 4.0) can have stricter timeliness and throughput requirements. Moreover, compared to regular cloud computing sites, edge sites are significantly smaller, which can result in increased workload volatility, as the balancing effect of the central limit theorem is related to size.

Containerization technologies enable *resource sharing* between applications via *logical isolation*. As a comparison, containers are similar to *virtual machines*, but generally, they are more resource efficient, because a larger part of the system is shared. However, despite of the nearly perfect logic isolation between different functions, *performance related interferences* can occur like the noisy neighbor phenomenon endangering temporal requirements.

For instance, *Kubernetes* (K8s) is an open-source framework for deploying, scaling and managing containerized applications on clusters of machines. The smallest unit in K8s is a *container* that encapsulates the run-able artifacts and their configurations. Containers are bundled into *pods*. Pods serve as the deployment units that are deployed on machines, called *nodes*. In K8s, machines can be both virtual and physical, allowing flexible hybrid setups. *Supervisory control* in K8s provides various data center styled services to improve extra functional properties, such as: load balancing, self-healing, automated scaling etc.

We focus on K8s in our examples, because technologies that can improve efficiency are essential to the edge and K8s is a prime candidate for that.

Motivated by an edge computing site where a large variety of failure modes must be anticipated with related resilience mechanisms, our motivating general question is: *how to select between the system-design alternatives to cover both functional and resource sharing originating faults?*

A *failure* is when the system cannot meet its requirements and deviates from its specification. The duration in the failed system state is *downtime*, while the duration in proper system state is *uptime*. We focus on three main aspects of the *failure-mitigation process*: availability, downtime, and risk.

Availability is the readiness for correct service measured as:

$$\text{Availability} = \frac{MTTF}{MTTF + MTTR}, \text{ where} \tag{1}$$

MTTF = mean time to failure = the expected uptime

MTTR = mean time to repair = the expected downtime

In Kubernetes, a repair typically involves a failover with reconfiguration time. Meanwhile, restarting a pod invalidates software state, thus might involve a time consuming re-initialisation process.

Downtime is the timeliness aspect of restoring a system state (*restoration time*): the time between fault activation and the completion of its mitigation. For example, when a Kubernetes node fails and the related pods are migrated to another node, then restoration finishes with the successful re-initialisation and restarting of the applications migrated within the pods.

Risk is defined in ISO 31000 as the "effect of uncertainty on objectives". Aligned with risk management, our analysis focuses on *per-failure-risks* (uncertainty of the failure-mitigation process) over a *mission duration*, because real life systems usually have availability and maximum downtime-per-failure requirements. This is especially true for critical CPSs with soft/hard real-time requirements and long mission/life times. For example, based on IEC 61508 the following *severity classes* can be inferred:

1. **Negligible**: minor injuries or less, loss less than $10K
2. **Marginal**: major injuries, loss exceeding $10K but less than $200K
3. **Critical**: single loss of life, loss exceeding $200K but less than $1M
4. **Catastrophic**: multiple loss of life, loss exceeding $1M

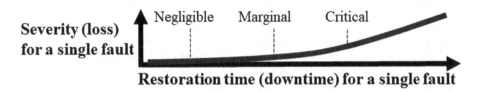

Fig. 1. Example: severity over downtime (restoration time) for a single fault

The severity of a fault depends in many systems on the downtime, as short interruptions are less dangerous than a longer loss of control. For our pilot application, we assume that the severity increases with downtime for a single fault (Fig. 1). Consequently, *we focus on the maximum downtime with the most severe consequences*, and to this end we employ *Extreme Value Analysis (EVA)* to handle rare, unusually large values manifesting in long-tailed distributions.

This paper is organized as follows: Sect. 2 presents EVA, Sect. 3 presents risk modeling, Sect. 4 presents a case study, Sect. 5 presents an extension of the method and the case study, and Sect. 6: conclusions.

2   Extreme Value Analysis

Critical applications need a special care of *extreme values* (long execution times, peak resource utilisation etc.) as they may lead to violations of extra-functional requirements, potentially resulting in critical operational situations.

Worst Case Execution Time (WCET) estimation is an established engineering field important in many areas, especially, soft/hard -real-time applications with timeliness constraints. WCET has two major paradigms [9]:

1. *Static Timing Analysis* (STA) is essentially a *white box* technique calculating execution times from component characteristics along a data-flow model of the system. Generally, complexity makes STA difficult [12].

2. *Measurement-Based-Timing-Analysis* (MBTA) is a *black box empirical evidence-based* approach, which can better handle complexity. *Measurement-Based Probabilistic Timing Analysis* (MBPTA) is a variant of MBTA which uses statistics to extract and fit a pWCET distribution as model over representative execution time data [3,11]. EVA is dominant for pWCET estimation, because its fundamentals make it suited for estimating the extreme domain, but using pWCET in a risk context is less elaborated [8]. Given a random variable X representing the execution time with its *cumulative distribution function* (*cdf*) $F_X(x)$ and probability density function (*pdf*) $f_X(x)$, the probability of a restoration time that remains below a threshold value x:

$$P(X \leq x) = F_X(x) = \int_{-\infty}^{x} f_X(x)dx \tag{2}$$

Meanwhile, the expected value of restoration time $E(X) = MTTR$.

pWCET is very effective in modeling simple systems of a sufficiently homogeneous behavior. However, many complex systems are multi-modal, frequently with a highly unbalanced occurrence of the different operation modes, like "normal" operation and "extreme" operation during recovery.

Proper statistical approximation is a common issue of modeling. In case of extreme values, the usually limited amount of "extreme data" owing to the rarity of the extreme values is overly dominated by "normal data". The *pdf*s of classical mono-modal distributions with typically bounded standard deviations, used in traditional pWCET, converge to zero too fast and thus suppress extremities.

The resulting high uncertainty in the domain of extreme values can lead to underestimation, and consequently in under-dimensioned systems. For instance, resiliency mechanisms with timeliness guarantees on downtime need faithful estimates of the extremely long repair times and associated risks as well.

Extreme value analysis is a branch of statistics dealing with extremely deviating values. This EVA introductory section is based on the following references: [10,17,19]. The classic case of EVA is in hydrology targeting both *analysis*: "What is the probability of an embankment surviving the floods of the next time period (e.g. 100 years)?" and *dimensioning*: "How tall embankment is needed to survive the floods of the next time-period with a given probability?". Analogously, a typical problem in an IT system is the sufficiency of slack resources for surviving rare peak workloads.

The starting point of EVA is a series of observations, which contains a (typically unbalanced) mixture of the "usual" values from the "normal" operation domain, and some "extreme" values from the "extreme" operating domain. EVA aims to fit a distribution which properly describes the extreme values as well.

The core of EVA is the separation of extreme values from the normal ones. The two main EVA algorithms are: *Annual Maxima Series* (AMS), a.k.a the block maxima; and *Peak Over Threshold* (POT), a.k.a threshold exceedance.

POT looks for a cutting threshold between the "normal" and "extreme" domains, and fits a Generalized Pareto Distribution (GPD) to the extremes.

AMS searches an ordered series of data (like a time-series) for large representative values by slicing the data set into equal length blocks, selecting only the maximum value from each block and discarding all the other values.

AMS uses standard *extreme value distributions* (EVDs) to fit these selected maxima. The three separate EVD classes are the Gumbel, Frechet, and Weibull distributions. These standard EVDs differ by the convergence rate (speed) of their density function's respective tail distributions.

1. The reference rate of convergence is the *exponential tailed Gumbel* family with example members: normal, gamma, log-normal, exponential.
2. The slowest is the *Fréchet* family with *heavy-fat tails* decreasing as a power function. Example class members: Pareto, Student, Cauchy, Burr.
3. The *Weibull* family has the *fastest tail convergence* with a finite right endpoint (thin tail). Example class members: uniform, beta and reverse Burr.

Block-size selection for AMS and threshold selection for POT are challenging, but there are several alternative methods available for these selections [7,20].

EVA modeling means selecting a candidate distribution function from the set of EVDs or GPD and parametrizing it for a best fit. The notion of the *maximum domain of attraction* (MDA) helps distribution fitting. Inside an MDA it is possible to setup convergence between a target distribution model and the distribution of observations of the maximum values of independent and identically distributed (*iid*) random variables using only normalising constants.

Essentially, MDA binds the EVD class members: the EVD class members can be different while retaining the same asymptotic tail behavior, thus giving an additional degree of freedom for fitting the normal and extreme domains.

POT estimates the distribution of the extreme values above a threshold u. The *conditional excess distribution* F_u over threshold u can be expressed as:

$$F_u(x) := P(X - u <= x | X > u) = \frac{F(u + x) - F(u)}{1 - F(u)} \tag{3}$$

$$0 <= x <= x_f - u$$

F: unknown *cdf* of random variable X

x_f: the finite or infinite right endpoint of the underlying distribution

The *conditional excess* has different names in different fields: *residual-life* in reliability and medical statistics, or *excess-of-loss* in insurance analytics.

This paper focuses on POT, because (1) it uses data more efficiently by selecting all the extreme value observations, (2) in engineering terms, it is comparable to the probability of slack resources exhaustion and timing threshold violation.

POT uses the family of GPD for distribution fitting and has several tuning parameters. The *location parameter* (μ) shifts the *pdf* along the X axis. The *scale parameter* (β) controls how spread out the *pdf* is along the X axis. The *shape parameter* (ξ) controls the behaviour of the distribution.

Based on the shape values (ξ), the distribution behaves the following way:

$$\xi \begin{cases} > 0 \text{ Pareto type I distribution with tail index } \alpha = \frac{1}{\xi} \\ = 0 \text{ exponential distribution} \\ < 0 \text{ Pareto type II distribution on a bounded interval } [0, -\frac{\beta}{\xi}] \end{cases} \quad (4)$$

The shape parameter has an important role in reliability engineering when GPD describes instantaneous failure rate.

$$\xi \begin{cases} > 0 \text{ decreasing failure rate} \\ = 0 \text{ constant failure rate, memory-less, evergreen} \\ < 0 \text{ increasing failure rate} \end{cases} \quad (5)$$

Note, that by characterizing the system behavior at the extremes, the type and shape of EVA distributions carry an important message regarding the underlying technology mechanisms. Owing to this, among other things, EVA could be used for statistical failure hypothesis analysis and anomaly detection.

For the sake of completeness, the GPD has the following *cdf*:

$$G_{(\xi,\beta)}(x) \begin{cases} 1 - (1 + \frac{\xi(x-\mu)}{\beta})^{\frac{-1}{\xi}} & \xi > 0, \beta > 0, x >= 0 \\ 1 - exp(\frac{-(x-\mu)}{\beta}) & \xi = 0, \beta > 0, x >= 0 \\ 1 - (1 + \frac{\xi(x-\mu)}{\beta})^{\frac{-1}{\xi}} & \xi < 0 < \beta, 0 < x <= -\frac{\beta}{\xi} \end{cases}, x \in \mathbb{R} \quad (6)$$

To summarize, if the maximum value of a metric is important for any reason (e.g. catastrophic consequences for underestimation), then EVA offers a fixed set of distributions to be fitted. If no EVD or GPD fits, then the extremity of the system under observation is out of the modelling power of EVA.

2.1   Detecting Extremity

Traditional statistics use dispersion or deviation as a measure of the amount of variation of a value. However, when rare values occur, they are hardly observable by these metrics, as they are suppressed by the majority of normal values. This necessitates the introduction of other statistical metrics specific to extreme values. Here we present two metrics that can be used as indicators for extremity. While a high indicator value implies extremity, a low indicator value provides only a statistical hint on their absence, but not an absolute guarantee.

Kurtosis is a measure of extremity expressing the long/heavy tailedness of a distribution. It is calculated based on the the fourth standardized moment. A higher kurtosis means more deviations, longer and heavier tails. As a reference, the kurtosis of a univariate normal distribution is 3, thus a kurtosis higher than 3 implies that the underlying process produces more frequent and/or more extreme values than the normal distribution.

Skewdness is a measure of "asymmetricity", expressing how weight is distributed compared to the center of mass (mean). A positive skew means that the left tail of the distribution has more mass compared to the right, resulting in a longer/thinner right tail towards high values. As a reference, the normal distribution has a skewdness of 0, because it is symmetric. Owing to this, a positive high skewdness implies that the underlying process produces more frequent and/or more extreme values than the normal distribution.

3 Modeling Risk

The previous section presented how EVA can extract a statistical distribution model based on observation data from systems manifesting occasional extremity.

The current section aims at downtime-related risk estimation. A method that can accurately capture the characteristics of extreme values is essential for downtime-related risk estimation, because in our focus domain of soft/hard real-time CPSs, the failure-related costs are significantly affected by downtime. Owing to this, the ability of EVA to properly estimate pWCET (extreme long restoration times related to the most severe consequences) is essential.

We assume a target system with a *long designated mission time* during which failures occur, and the built-in supervisory system initiates an automatic mitigation with an associated cost. The system has a total budget of restoration costs for the mission time, and budget exhaustion means a complete system failure.

With our focus on critical systems, the cost function has to represent several severity classes. We define the cost as a function of the restoration time (downtime), expressing the time-dependent impacts of a failure.

Based on IT systems literature, where linear utility functions are used [1, 21], we assume for simplicity of the analysis that the *cost function* $CF(t)$ is a continuous, monotonic, non-decreasing, invertible, differentiable and integrable function of the restoration time t. The *inverse cost of failure* $CF^{-1}(c)$ maps a given cost to the corresponding downtime.

The *probability of a failure-related cost being under a value (c)* is defined, as usual, by a *cdf*: $P_{CF}(C \leq c) = F_X(CF^{-1}(c))$. EVA contributes to calculating the cost *cdf* through the estimation of $F_X(x)$ representing the restoration time.

3.1 Probabilistic Modeling for a Mission Duration

We assume the failures over a mission to be independent because (1) rigorous checking of critical applications prior deployment eliminates (the majority of) systematic failures, making random failures dominant; (2) the MTTR in such IT systems is orders of magnitude smaller than MTTF due to automated recovery.

Note, that this last restriction is not very strict. EVA is still viable if the number of failures is a *sum of correlated failures* due to a generalization of the central limit theorem to non-independent random variables using extreme value statistics [4]. Thus EVA tolerates some correlation when estimating the sums.

We define a random variable Y representing the *number of failures* during a mission with *cdf* $F_y(y)$ and *pdf* $f_y(y)$. Similar to WCET estimation, EVA can be used for a worst case like estimation of $F_y(y)$.

3.2   Mission Risk: The Cumulative Cost of Failures

The expected number of failures $E(Y)$ and the related expected cumulative cost of failures $E_{CFT}(t)$ for a mission time t can be estimated as:

$$E(Y) = \int_{-\infty}^{\infty} y f_y(y) dy = \text{t *Failure rate} = t \frac{1}{MTTF} = \frac{t}{MTTF} \qquad (7)$$

$$E_{CFT}(t) = \frac{t}{MTTF} \int_{-\infty}^{\infty} CF(x) f_X(x) dx, \qquad (8)$$

However, expected values alone are insufficient characteristics of a critical system, thus the current section addresses the worst case scenario of mission failures.

A *mission failure* occurs if the *cumulative cost of failures* (CCF) exceeds the pre-allocated *mission budget* (b). *Mission risk* is the probability of budget exceedance: $P(CCF > b)$ given (1) the distribution of failures during a mission $F_y(y)$, (2) the cost function CF, and (3) the distribution of the restoration time $F_X(x)$ (the fitted pWCET curve from EVA). The probability of budget exceedance is analogous to the aggregate claim amount in a risk-insurance context.

Calculating the probability of budget exceedance is far from trivial. As a consequence, we will briefly highlight some alternative methods for calculation.

(1) *Pricing the Observations*: If the complete series of downtime logs is available, their cost can be calculated on an occurrence-by-occurrence basis. EVA or any other estimator could be used to estimate the probabilistic properties of the mission risk by analyzing the individual and cumulative costs.
(2) *Simulation:* if data is not directly available, but the essential components of a system are available, such as a fitted pWCET curve output of EVA, then the known components can be used to simulate data. After that, techniques based on data can be used, similar to the previous point.
(3) *Analytical Methods* [16]: if the properties of related random variables are known, such as a fitted pWCET curve output of EVA, then mathematical methods can be used to calculate the probability of budget exceedance. For discrete random variables, with known probability mass functions (*pmf*), the exact *pmf* of the sum can be calculated using *discrete convolutions*. If the *pmf* takes only non-negative integer values, then a power series representation of the *pmf*, known as *probability-generating function*, could be used, where the sum becomes a product of the generating functions. For continuous random variables, with known *pdfs*, the exact *pdf* of the sum can be calculated using *continuous convolutions*. However, calculating continuous convolutions is difficult, because often there is not always a clean closed

form. For GPD, a complex formula exists for the continuous convolution [18].

(4) *Approximation:* the *Feller convolution theorem* is applicable for the estimation of the convolution of heavy/long tailed distributions [13], where the sum of the random variables can be approximated by the value of a single large value. This is especially applicable for CPSs, where one severe failure can dominate multiple less severe ones.

3.3 Workflow

Our workflow has two main blocks: (1) *Measurement-based EVA*, and (2) *risk model and availability evaluation* (Fig. 2). The Measurement-based EVA block contains the measurements and the data analytics. Data analytics involves fitting *cdf*-s, where EVA is a component that can be used for long/heavy tailed distributions, but other *cdf* estimators can be used as well. This block is equivalent to WCET estimation using MBPTA in case of execution time measurements, such as restoration time. The risk model and availability evaluation block contains *Design Space Exploration* (DSE) over the risk model components and the risk calculator. The risk calculator calculates the mission risk (the cumulative cost of failures) using any method from the previously presented ones or other methods.

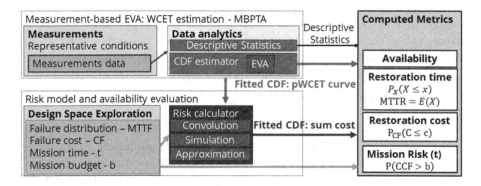

Fig. 2. Workflow: measurements, analytics, risk calculation, design space exploration

Our workflow has the following phases:

1. Measurement-based EVA
   (a) Acquiring representative measurement data using different measurement configuration parameter setups (e.g. measuring restoration time).
   (b) Data analytics for the extraction of statistical models, like *cdf*s, and the execution of EVA (e.g. fitting the pWCET curve).
2. Risk model and availability evaluation
   (a) Statistical model selection and parametrization for the risk model components using domain knowledge. This step involves a DSE like parameter sweep (e.g. failure distribution, MTTF, mission time).

(b) Risk calculation using the risk model components (e.g. MTTT, mission time) and the *cdf*s from the measurements (e.g. fitted pWCET curve).
(c) Computation of other output metrics (e.g. availability)

We implemented the simulator for a single mission the following way:

1. Parametrisation of $F_y(y)$(the failure distribution for number of failures) with MTTF and mission time based on Eq. (7).
2. Determination of the number of failures (NF) during the mission by drawing a rounded random number from the parametrised $F_y(y)$ for the mission.
3. Determination of the downtime of each failure by drawing a random number from the empirical $F_X(x)$(downtime, restoration time) for each failure.
4. Replacement of the downtime of each failure that was above the EVA threshold with a new random number drawn from the previously fitted GPD (EVA fitted $F_X(x)$). This step is needed because GPD models the conditional excess distribution: $F_u(x) := P(X - u <= x|X > u)$. Thus, this is the key to proper long tail modeling, because by replacing the over the threshold values with EVA values, we essentially eliminate the potential under estimation of the empirical distribution. The empirical distribution is only needed to ensure the ratio of over the threshold and under the threshold values.
5. Calculation of the cost of downtime for each failure using the cost functions.
6. Calculation of the mission cost by summing the cost of each failure.

Note that WCET only serves as an example in the presented workflow, because any other metric could be used, for example the peak (worst case) workload from a dimensioning (resource planning)/capacity design context [6].

4   Case Study

The subject of the case study was a typical element of the edge: Kubernetes. We selected model input parameters for the DSE like parameter sweep to be reasonable within the context of soft/hard-real-time CPSs running on the edge.

For calculating the mission risk we used simulations, as described before, because other methods were difficult to implement or computationally complex.

We used R for data analytics and simulation. For EVA POT threshold selection we used the parameter stability plot function "gpd.fitrange" [14], and mean residual life plot function "mrlplot" [15]. For fitting the GPD, we used maximum likelihood estimation implemented in "fevd" [15]. For validating the fitted model, we used QQ plot, empirical and modeled density plot, and goodness of fit tests (GOF) with significance level 0.05 implemented in "gpdSeqTests" [2].

4.1   Measurement-Based Extreme Value Analysis

We used official Kubernetes over a *Dell PowerEdge R510* as experimental platform. The configuration under evaluation contained two machines: (1) an *application emulator* running a synthetic benchmark of parametrizable utilization for its different resources (2) the *orchestrator* performing supervisory control, instrumentation and logging of the measurement campaigns, and controlling fault mitigation. The later one incorporated a simple script executing software implemented fault injection (SWIFI) via the ''`docker kill`'' command.

We defined downtime as the time between *fault injection* and container *restart completion*. Due to our focus on the fundamental mitigation services of the infrastructure, our definition excludes application dependent initialisation.

In a previous experiment we focused on executing an MBPTA based WCET estimation using EVA on the same K8s setup *without any significant* workload [5]. Surprisingly, we found in this simplest case the presence of container restart time extremity in a nearly idle system. Moreover, EVA proved to be a more robust estimator to the extreme values than a naïve non-EVA estimator.

In the current series of fault injection campaigns, we conducted measurements with different levels of resource utilization to investigate their effects. A total of 12 different (CPU [%], Disk IO [%]) background utilisation pairs were used in the different measurement campaigns: (0,0), (50,0), (80,0), (95,0), (0,5), (50,5), (80,5), (95,5), (0,40), (50,40), (80,40), (95,40); the pairs are expressed in percentage. Each measurement campaign had 1000 measurements. Based on the data, we created 12 separate system characteristics and pWCET curves.

Table 1. Workload based container restart time statistics with GPD goodness of fit

Configuration		Container restart time					POT	GOF
CPU[%]	DiskIO[%]	Min[s]	Mean[s]	Max[s]	Skew.	Kurt.	Threshold	Pvalue
0	0	1.9	2.9	3.9	−0.5	4.7	3.08	0.93
0	5	2.7	4.0	5.9	0.3	3.0	5.38	0.64
0	40	3.8	13.6	30.4	1.2	4.5	20.80	0.93
50	0	2.2	3.1	4.0	−0.3	3.8	3.48	0.07
50	5	2.9	4.3	6.1	0.2	2.9	5.15	0.79
50	40	4.3	14.0	32.1	1.1	5.1	12.30	0.43
80	0	2.3	3.3	4.3	0.0	2.5	3.75	0.57
80	5	3.3	4.8	6.0	−0.2	2.8	5.18	0.93
80	40	4.6	14.2	34.2	1.3	6.1	15.55	0.21
95	0	2.7	3.9	5.1	0.3	2.8	4.10	0.64
95	5	3.4	5.1	6.8	0.1	2.8	5.55	0.93
95	40	4.6	15.0	35.4	1.5	6.8	18.90	0.29

Based on measurements (Table 1), the container restart process is significantly affected by background load, both in the normal (mean) and the extreme domains (maximum). The restart time is more sensitive to DiskIO than CPU load. Regarding restart time Max, Kurtosis and Skewdness, it can be concluded that extremity is present in the system and it increases with the workload.

A practical finding: for a fast container restart, the DiskIO has to be kept low, e.g. under 5%, implying a DiskIO overhead factor of 19×. However, avoiding the critical impact of extreme values by such a pure dimensioning may kill a main advantage of edge computing: a good utilization via resource sharing.

4.2 Risk Model and Availability Evaluation

We selected 10^7 s (115.74 days) as MTTF to represent thoroughly tested components producing only a few failures per year on average. Additionally, we chose a smaller MTTF value of 10^5 s (1.1574 days) to represent a serial system of 100 components, which is realistic for a microservice architecture based edge cloud-native services with dozens of containers. Note that selecting a particular system architecture does not confine the generality of the approach.

For modeling the failure distribution over a mission, we used the exponential distribution to represent a memory-less system with a constant failure rate.

For domain-wise modeling of the restart times, we used the empirical distribution for the "normal" domain, while using EVA-GPD for the "extreme".

We ran 100000 simulations per configuration for the mission cost calculation.

For the cost of a single restoration, we defined hypothetical time intervals with increasing associated fix costs to represent 3 different levels of severity. Meanwhile, we considered restart times under 3 s free.

For mission times, we used years as units, to represent longer infrastructure investment lifetimes, typical for an industrial context.

We also defined seven hypothetical mission budget values.

To summarize, for each of the 12 measurement configurations, we explored $2 * 3 * 7 * 3$ model variants, resulting in a total of $12 * 2 * 3 * 7 * 3 = 1512$ evaluations:

- 12 measurement configurations regarding (CPU[%] & DiskIO[%]) pairs
- 2 MTTFs [seconds]: 100000.0 (1.1574 days), 10000000.0 (115.74 days)
- 3 mission times [years]: 1, 2, 4
- 7 mission budgets [USD]: 10000, 20000, 30000, 40000, 80000, 160000, 320000
- 3 cost functions:
  - soft (0–3 s: 0 USD), (3–6 s: 100 USD), (6–12 s: 1000 USD), (12–24 s: 10000 USD), (24- s: 100000 USD)
  - hard (0–3 s: 0 USD), (3–6 s: 1000 USD), (6–12 s: 10000 USD), (12–24 s: 100000 USD), (24- s: 1000000 USD)
  - extra hard (0–3 s: 0 USD), (3–6 s: 10000 USD), (6–12 s: 100000 USD), (12–24 s: 1000000 USD), (24- s: 10000000 USD)

Looking at the availability and the maximum simulated mission costs (Fig. 3), it can be concluded that as availability decreases, the maximum mission cost increases in a superlinear way due to increasing cost penalty on long downtimes. This effect is further amplified if severe cost functions harshly penalize the longer restart times, which are common in critical CPS-s.

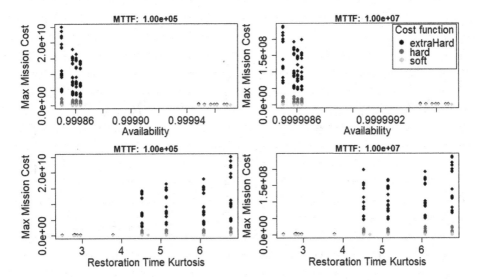

Fig. 3. Max. mission cost over availability and long tailedness of restoration time measured by kurtosis for different cost functions and MTTFs

However, the mission risk can be significantly different for the same level of availability, especially for lower availability or longer mission times. This implies that as the "normal" domain dominates availability through the MTTR, the "extreme" domain dominates the mission risk through the cost of very large restoration times. This emphasises the necessity to asses the "extreme" domain.

Looking at the kurtosis of restoration times and the maximum simulated mission costs (Fig. 3), a similar conclusion can be drawn: extremity in the restoration time radically increases risk, indicated by very large maximum simulated mission costs; making restart time kurtosis a viable indicator for mission risk.

5   Extension to Cold-Backup

This section presents how a cold backup redundancy scheme can be assessed using the the proposed workflow. When a restart is initiated a spare container is also started and the faster to restart would become the active container. In K8s, it can be implemented as a spare Docker container laying dormant on a node.

We modified the simulator accordingly: we calculate two simulated downtime values for each failure and select the smaller value (step 3 and 4 executed twice).

Looking at the maximum mission costs of the cold backup results (Fig. 4), it can be concluded that the maximum simulated mission costs for cold backup were visibly smaller. By comparing the distributions of max. mission cost (Fig. 5), they appear similar. For MTTF 10^5 a p-value of 0.064 for Kolmogorov-Smirnov (KS) test indicates they are the same. Meanwhile for MTTF 10^7 a KS p-value of 0.009 indicates otherwise. This means that the cold backup had a larger impact

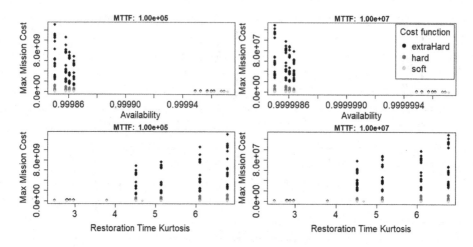

Fig. 4. Cold backup: maximum mission over other parameters

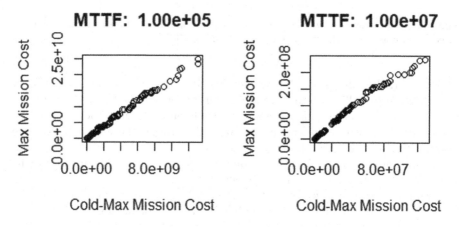

Fig. 5. QQ plots: comparing distributions of Max. mission costs

when the number of failures was small. This is expected, as the probability of extreme long restoration changes from $P(X \geq x)$ to $P(X \geq x)^2$ for cold backup.

Based on the results, it can be concluded that a cold backup redundancy is a viable risk mitigation strategy with effectiveness depending on the ratio of normal restoration costs versus extremes within the mission cost.

6   Conclusion

We devised a method that augments WCET with risk and uses DSE. In the current evaluation, we compared dimensioning and cold backup alternatives. The evaluation methodology is universal and adaptable, because it can accommodate a variety of cost functions and random variables.

Based on the results, we found that in K8s (1) container restart times extremity increases risk, (2) workload increases container restart extremity, (3) harsher cost functions increase risk, (4) dimensioning is costly for risk mitigation, and (5) cold backup redundancy is viable for risk mitigation.

References

1. Ardagna, D., Trubian, M., Zhang, L.: SLA based resource allocation policies in autonomic environments. J. Parallel Distrib. Comput. **67**(3), 259–270 (2007)
2. Bader, B., Yan, J.: eva: Extreme Value Analysis with Goodness-of-Fit Testing (2020)
3. Bernat, G., Colin, A., Petters, S.M.: WCET analysis of probabilistic hard real-time systems. In: 23rd IEEE Real-Time Systems Symposium, RTSS 2002, pp. 279–288 (2002)
4. Bertin, E., Clusel, M.: Generalized extreme value statistics and sum of correlated variables. J. Phys. A Math. Gen. **39**(24), 7607–7619 (2006)
5. Bozóki, S., et al.: Application of extreme value analysis for characterizing the execution time of resilience supporting mechanisms in kubernetes. In: Bernardi, S., et al. (eds.) EDCC 2020. CCIS, vol. 1279, pp. 185–199. Springer, Cham (2020). https://doi.org/10.1007/978-3-030-58462-7_15
6. Bozoki, S., Pataricza, A.: Extreme value analysis for capacity design. Int. J. Cloud Comput. **7**(3–4), 204–225 (2018)
7. Caeiro, F., Gomes, M.: Threshold Selection in Extreme Value Analysis: Methods and Applications, pp. 69–86. Taylor & Francis (2015)
8. Castillo, E., Hadi, A., Balakrishnan, N., Sarabia, J.: Extreme Value and Related Models with Applications in Engineering and Science. Wiley, New York, USA (2004)
9. Cazorla, F.J., Kosmidis, L., Mezzetti, E., Hernandez, C., Abella, J., Vardanega, T.: Probabilistic worst-case timing analysis: taxonomy and comprehensive survey. ACM Comput. Surv. **52**(1), 1–35 (2019)
10. Cizek, P., Härdle, W.K., Weron, R.: Statistical Tools for Finance and Insurance. Springer, Berlin, Heidelberg (2011)
11. Cucu-Grosjean, L., et al.: Measurement-based probabilistic timing analysis for multi-path programs. In: 2012 24th Euromicro Conference on Real-Time Systems, pp. 91–101 (2012)
12. Cullmann, C., et al.: Predictability considerations in the design of multi-core embedded systems. In: Proceedings of Embedded Real Time Software and Systems, pp. 36–42 (2010)
13. Feller, W.: An introduction to probability theory and its applications. John Wiley & Sons, New York, USA (2008)
14. Gilleland, E., Heffernan, J.E., Stephenson, A.G.: ismev: An Introduction to Statistical Modeling of Extreme Values (2018)
15. Gilleland, E., Katz, R.W.: extRemes 2.0: an extreme value analysis package in R. J. Stat. Softw **72**(8), 1–39 (2016)
16. Grinstead, C.M., Snell, J.L.: Introduction to probability. Am. Math. Soc. (2012)
17. McNeil, A.J., Frey, R., Embrechts, P.: Quantitative Risk Management: Concepts Techniques and Tools - Revised Edition, Revised edn. Princeton University Press, New Jersey, USA (2015)

18. Nadarajah, S., Zhang, Y., Pogány, T.K.: On sums of independent generalized pareto random variables with applications to insurance and cat bonds. Probab. Eng. Inf. Sci. **32**(2), 296–305 (2018)
19. Rakoncai, P.: On Modeling and Prediction of Multivariate Extremes. Ph.D. Thesis, Mathematical Statistics Centre for Mathematical Sciences, Lund University (2009)
20. Scarrott, C., MacDonald, A.: A review of extreme value threshold estimation and uncertainty quantification. Revstat Stat. J. **10**, 33–60 (2012)
21. Zhang, L., Ardagna, D.: SLA based profit optimization in autonomic computing systems. In: Proceedings of the 2nd International Conference on Service Oriented Computing, pp. 173–182 (2004)

Evaluation Framework for Performance Limitation of Autonomous Systems Under Sensor Attack

Koichi Shimizu[1]([⊠]), Daisuke Suzuki[1,4], Ryo Muramatsu[1], Hisashi Mori[1], Tomoyuki Nagatsuka[2], and Tsutomu Matsumoto[3,4]

[1] Mitsubishi Electric, Kanagawa, Japan
`shimizu.koichi@ea.mitsubishielectric.co.jp`
[2] Mitsubishi Electric Engineering, Kanagawa, Japan
[3] Yokohama National University, Kanagawa, Japan
[4] National Institute of Advanced Industrial Science and Technology, Tokyo, Japan

Abstract. Autonomous systems such as self-driving cars rely on sensors to perceive the surrounding world. Measures must be taken against attacks on sensors, which have been a hot topic in the last few years. For that goal one must first evaluate how sensor attacks affect the system, i.e. which part or whole of the system will fail if some of the built-in sensors are compromised, or will keep safe, etc. Among the relevant safety standards, ISO/PAS 21448 addresses the safety of road vehicles taking into account the performance limitations of sensors, but leaves security aspects out of scope. On the other hand, ISO/SAE 21434 addresses the security perspective during the development process of vehicular systems, but not specific threats such as sensor attacks. As a result the safety of autonomous systems under sensor attack is yet to be addressed. In this paper we propose a framework that combines safety analysis for scenario identification, and scenario-based simulation with sensor attack models embedded. Given an autonomous system model, we identify hazard scenarios caused by sensor attacks, and evaluate the performance limitations in the scenarios. We report on a prototype simulator for autonomous vehicles with radar, cameras and LiDAR along with attack models against the sensors. Our experiments show that our framework can evaluate how the system safety changes as parameters of the attacks and the sensors vary.

Keywords: Autonomous systems · Safety · Security · Sensor attack · SOTIF · Performance limitation · STAMP/STPA

1 Introduction

Autonomous systems such as autonomous vehicles rely on various sensors to perceive the surrounding world and decide what to do next. There have been a lot of reports on attacks against sensors, e.g. magnetic wheel speed sensors [1],

I. Habli et al. (Eds.): SAFECOMP 2021, LNCS 12852, pp. 67–81, 2021.
https://doi.org/10.1007/978-3-030-83903-1_1

gyro sensors [2], FMCW radar [3,4], and LiDAR [5,6], and against sensor-based autonomous systems [7]. The safety of autonomous systems against sensor attacks must therefore be assured. As an illustrative example, we use an AEB-equipped car with a radar, a camera and a LiDAR (Fig. 1) throughout the paper. AEB (Autonomous Emergency Braking) uses the sensors to detect objects around the car, and measure the distance to and relative speed of the nearest one in front. If it detects an impending crash, it will dispatch a warning or apply the brakes. There is high risk of serious accidents if the sensors are compromised.

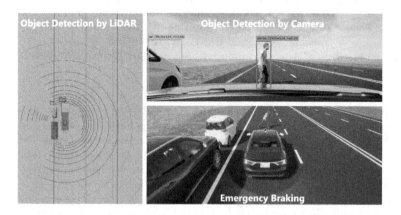

Fig. 1. AEB equipped-car with a radar, a camera and a LiDAR

To assure the safety of autonomous systems, scenario-based simulation [8,9] is widely accepted as a key tool because real-world testing for hundreds of millions of miles [10] is unrealistic. One of the issues of scenario-based simulations is how to select a set of relevant scenarios from the vast space of scenarios consisting of many parameters. In addition, there has been few autonomous system simulator that embeds sensor attack models as far as we know.

In this paper we propose a framework to evaluate performance limitations of autonomous systems in the light of SOTIF. It combines STAMP/STPA-based safety analysis to identify sensor attack scenarios to be evaluated, and sensor attack simulation to evaluate the effect of sensor attacks in the scenarios. We elaborate on safety analysis steps and results for AEB-equipped cars, and provide a prototype of a sensor attack simulator and examples of evaluation using it.

Contributions. The main contributions of this paper are threefold:

- Evaluation framework of performance limitations that combines safety analysis and sensor attack simulation (Sect. 2).
- Method of attack scenario identification based on STAMP/STPA safety analysis together with concrete results for AEB (Sect. 3).
- Autonomous system simulator with sensor attack models embedded, and a prototype for AEB together with evaluation examples (Sect. 4).

2   Evaluation Framework Based on SOTIF Process

2.1   Relevant Standards and SOTIF

We mention two safety standards and one security standard in relation to our problem. ISO 26262 [11] and ISO/PAS 21448 [12] are safety standards for road vehicles. The former addresses functional safety as the absence of unreasonable risks caused by failures; The latter complements functional safety, addressing SOTIF (Safety Of The Intended Functionality) as the absence of unreasonable risks due to intended functionality or performance limitation. SOTIF takes into account sensors that advanced functionalities these days rely on. ISO/SAE 21434 [13] addresses the security aspects of road vehicles. It focuses on security risk management during the development process, and specific attacks are out of scope.

The notion of performance limitation in SOTIF with sensors in mind matches the purpose of evaluating how sensor attacks affect the system, e.g. which part or whole of the system will fail, or will keep safe, if some of the built-in sensors are compromised. Figure 2 gives an overview of the improvement process of SOTIF. Figure 2 depicts a cycle process in which Functional and System Specification is the starting point, hazard scenarios are identified for it, and functions are modified to mitigate the hazard factors. Performance requirements for sensors are thereby defined at the design stage. On the other hand, model-based design is widely accepted for autonomous systems such as vehicles and robots. It helps evaluate and improve the specification in a continuous manner from the early stages of development by means of a simulatable specification (called a model) throughout development. That suits the cycle process in Fig. 2. We therefore adopt a model-based design framework.

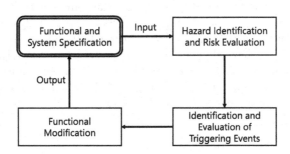

Fig. 2. SOTIF improvement process

2.2   Evaluation Framework

We present an evaluation framework for performance limitation under sensor attacks (Fig. 3). Aside from the framework itself, it is novel in two respects.

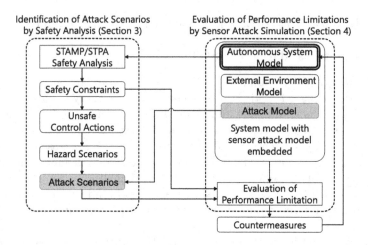

Fig. 3. Evaluation framework for performance limitation under sensor attacks

Firstly it identifies sensor attack scenarios by combining STAMP/STPA-based safety analysis with the knowledge of sensor attacks (Sect. 3). Secondly it embeds sensor attack models in a target autonomous system model to realize sensor attack simulation at system level. It is thereby able to evaluate the performance limitations of the target autonomous system in the attack scenarios (Sect. 4).

3 Identifying Attack Scenarios Using STAMP/STPA

3.1 STAMP/STPA Safety Analysis

Safety analysis is a tool to identify scenarios that can lead to hazards. Examples of safety analysis methods include FTA (Fault Tree Analysis) [14], FMEA (Failure Mode and Effect Analysis) [15], and STAMP/STPA [16]. While FTA and FMEA focus on hazards caused by component failures, STAMP/STPA[1] takes the view that hazards can also occur as a result of unintended interactions between components even if none of them has any failure. The view is compatible with SOTIF, and we therefore use STAMP/STPA.

3.2 Analysis Steps and Results

For the input Autonomous System Model in Fig. 3, we use the prototype of an AEB-equipped car (Sect. 4), from which we extract a control and feedback structure to be analyzed. Figure 4 shows the extracted structure that consists of the fewest components possible for brevity, e.g. sensors are not separated from AEB ECU. The labels at the bottom indicate the correspondence to Fig. 5.

---

[1] STAMP (Systems Theoretic Accident Model and Processes) is an accident causality model based on system theory, which underpins the analysis method STPA (System-Theoretic Process Analysis).

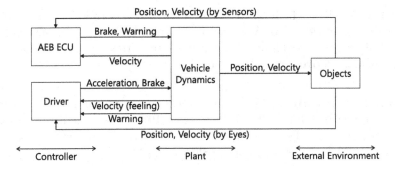

Fig. 4. Control and feedback structure of the target AEB-equipped vehicle

Safety Constraints. We first define safety constraints as the inverse of hazards that can lead to losses (e.g. injury, a loss of life, etc.). For the current example, we define five safety constraints shown in Table 1. We input them as evaluation criteria for performance limitation (see Fig. 5).

Table 1. Safety constraints

	Safety Constraint
SC1	Brakes applied in time to decelerate and stop ego vehicle if target object is within defined distance.
SC2	AEB emits warning sound to driver.
SC3	Ego vehicle stops if it crashes into object.
SC4	Deceleration or warning sound by AEB does not panic driver into mishandling ego vehicle.
SC5	AEB does not decelerate ego vehicle if there is no impending crash.

Unsafe Control Actions. The next step is to identify UCAs (Unsafe Control Actions) that can break the safety constraints. This is done in a systematic manner according to how control actions are applied: 1) providing, 2) not providing, 3) too early, too late, and 4) stopped too soon, applied too long. For the current example, we identify 21 UCAs, 14 of which are relevant to safety. For lack of space, we show only four UCAs in Table 2.

Table 2. Examples of identified UCAs

Control Action	Unsafe Control Action			
	Providing	Not providing	Too early, too late	Stopped too soon, applied too long
Brakes	Brakes are applied and ego vehicle decelerates and stops in spite of there being no object in front. (SC5)	Brakes are not applied and ego vehicle crashes into object in front. (SC1)	Brakes are applied too late and ego vehicle crashes into object in front. (SC1)	Brakes are applied too long and ego vehicle cannot restart after having safely stopped. (Not relevant to safety)

Hazard Scenarios. The final step of STAMP/STPA is to identify hazard scenarios for each of the UCAs. We identify 15 scenarios for the current example. Due to lack of space, we show only one of the scenarios together with example attack scenarios derived from it in Table 3.

Attack Scenarios. By linking sensor attacks to the causes of the hazard scenarios identified by STAMP/STPA, we create 102 attack scenarios in total for the current example. Table. 3 shows example scenarios that causes the "Not providing" UCA in Table 2, which can then break SC1.

Table 3. Examples of identified attack scenarios

Hazard Scenario	Sensor	Attack Type	Attack Scenario
AEB does not work due to sensor failure	Camera	Adversarial patch against single object detection	Camera fails to detect pedestrian due to object detection interference by adversarial patch. AEB does not work due to incorrect output of camera.
		...	...
	Radar	Denial jamming	Radar fails to detect pedestrian due to denial Jamming. AEB does not work due to incorrect output of radar.
		...	...
...	...	...	...

←————————→ ←————————————————————————————————————→
　By STAMP/STPA　　　　By linking hazard scenarios and sensor attack types

As to sensor attack types, we gather a list from existing works to cover the attack goals: interference that prevents object detection, and deception that modifies detection results. We then arrange them into 11 types according to the detailed methods or effects (Table 4). As of now, eight out of 11 (indicated by bullets in Table 4) are modelled for simulation while all of them are supported in scenario creation.

Table 4. List of sensor attack types

Target	Interference	Deception
Radar	•Denial Jamming [7,17]	•Range deception [3,7,17–19], •Velocity deception [17], False target jamming
Camera	•Adversarial patch against single/all object detection [20], Road marking modification	•Projection of false pedestrian or vehicle
LiDAR	•Light absorption [5], •Light injection [6],	•Spoofing by light injection [5]

4   Evaluating Performance Limitations Under Sensor Attacks

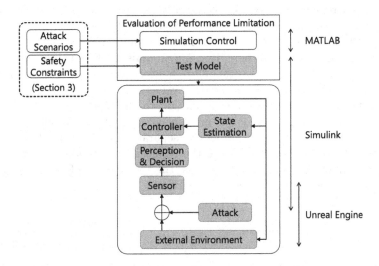

Fig. 5. Top-level structure of our sensor attack simulator

We present a sensor attack simulator to realize the right side of the framework in Fig. 3. The top-level structure is shown in Fig. 5. Given the attack scenarios and safety constraints, it evaluates the performance limitations of the target autonomous system by testing if the system satisfies the safety constraints in the attack scenarios. We choose MATLAB [21]/Simulink [22] as a platform widely used in model-based design together with Unreal Engine [23] to implement the external environment and its boundaries with sensors and attacks.

We build a prototype of an AEB-equipped car with a radar, a camera and a LiDAR based on a mix of two example models supplied by MathWorks, one for lane keeping assistance [24] and the other for AEB [25], together with necessary additions and modifications such as a LiDAR [26], object detection by YOLOv2 (You Only Look Once) [27], and fine-grained simulation of radar [28]. Sensor attack models are our own.

4.1   Test Model for Verification of Safety Constraints

The simulator must check if the target system satisfies the safety constraints, and if not, stop running. There are largely two methods for such evaluation: conventional testing and formal verification. They have their merits and demerits, and do not exclude but complement each other [29,30]. For example, formal verification can give a proof for the verification result by checking all possible states, while it can also lead to state explosion as the complexity of a system increases. One usage is therefore to formally verify the safety-critical part of the system and to test the system as a whole in a conventional way. In this paper we use a conventional testing method with the focus on evaluating the safety of the autonomous system as a whole.

For the current example of AEB, the safety constraint SC1 states that the AEB control is correct, which can be evaluated as follows: The target model maintains the positions and velocities of objects measured by sensors, and the AEB control calculated from them. It also maintains the true values of positions and velocities, and we can use them to calculate the true AEB control. By comparing the two AEB controls, we can evaluate if SC1 is met.

When we add a model for the evaluation to the simulator, it is desirable to keep the target system model as unchanged as possible. Simulink Test [31] has a mechanism called a test harness to separate the model for testing from the model under test. Figure 6 shows the resultant test model for SC1 in our prototype. Implemented as a test harness, the test model refers to and copies from the target model, but never changes it.

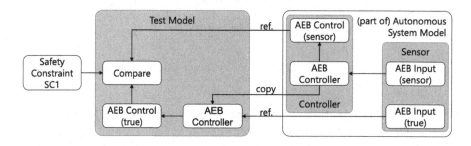

Fig. 6. Prototype model for testing the safety constraint SC1

4.2   Sensor Attack Simulator

The main body of the sensor attack simulator consists of seven models described below. All but the Attack model are assumed to be created through the development of the target system.

Plant, Controller, State Estimation. Those are the core of a control system. Our prototype is built around the vehicle dynamics (Plant), AEB controller (Controller and State Estimation) and other peripheral models provided by MathWorks.

Sensor, Perception & Decision, Attack. Those are to be designed considering what types of sensor attacks we want to evaluate. This paper aims to evaluate attacks on sensors on their own, on sensor fusion, and on signal processing. Therefore, the Sensor model is designed to include sensor fusion as well as separate sensors. Sensor fusion is further divided into two stages: detection concatenation and multi-object tracking. The Perception & Decision model is designed to include object detection algorithms CFAR (Constant False Alarm Rate) for the radar and YOLOv2 [32] for the camera. The Attack model considers those algorithms as well as the sensors on their own. The resulting models of Sensor, Perception & Decision and Attack are shown in Fig. 7.

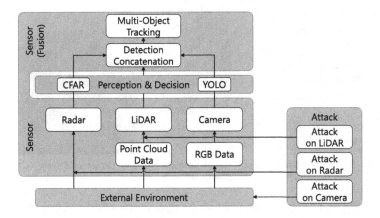

Fig. 7. Prototype models of sensor, perception & decision and attack

Our collection of attack models are as summarized in Table 4. We show examples of sensor attack simulation supported by the prototype in Fig. 8.

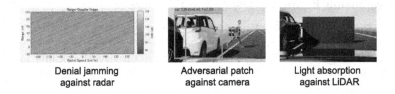

| Denial jamming against radar | Adversarial patch against camera | Light absorption against LiDAR |

Fig. 8. Examples of sensor attack simulation supported by the prototype

External Environment. It models the external environment surrounding the target system, e.g. nearby objects, how they are perceived by the sensors, and the positional relationship between the target system and the other objects. It also defines the temporal development of the target system and the environment as operational scenarios.

Our prototype models the external environment that complies with the evaluation criteria of AEB in JNCAP [33] and Euro NCAP [34], and supports the complete set of operational scenarios: largely, five scenarios of car detection and 11 scenarios of pedestrian and cyclist detection, and a total of 278 scenarios with parameter variations. As an example, Fig. 9 shows the CPNO (Car-to-Pedestrian Nearside Obstructed) scenario in JNCAP, where the ego vehicle travels forward towards a pedestrian crossing its path from the nearside who is out of sight at first due to stationary vehicles in between.

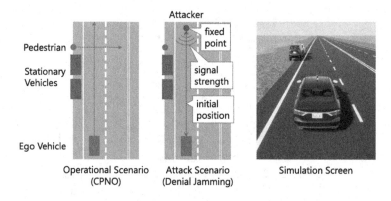

Fig. 9. CPNO scenario and an example of denial jamming in it

Figure 9 also shows an example attack scenario of denial jamming, which is taken from Table. 3 and embedded in CPNO by specifying the attack settings as follows: denial jamming is applied from a fixed point in front of the vehicle with the attacker's initial position and signal strength being variable.

4.3   Evaluation Examples Using the Prototype

We show three examples of evaluation using the prototype: one about attack parameters, and two about sensor design parameters. We use the attack scenario of denial jamming in CPNO (Fig. 9). For the sake of brevity, the simulation stops when the car crashes into the pedestrian instead of when the safety constraints such as SC1 are not met.

Jamming Attack on the Radar. We evaluate the effect of jamming attack on the radar with respect to two parameters: the attacker's position and signal strength. The other parameters are fixed: the velocity of the ego vehicle is 25 [km/h], and the signal strength of the ego vehicle, 10 [dBm]. The camera and the LiDAR are turned off in this evaluation.

Figure 10 shows the results. Each element of the matrix denotes whether the car crashes into the pedestrian (Crash) or not (Safe). The result is as expected: the stronger the attacker's signal is, or the nearer the attacker's position is, the more likely the attack is to succeed. That proves the validity of the simulation.

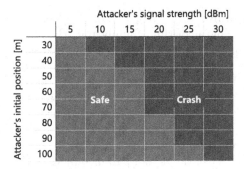

Fig. 10. Evaluation result with respect to radar jamming parameters. The camera and the LiDAR are turned off.

Detection Concatenation. As an example of sensor fusion, we evaluate the effect of concatenation of the radar and camera. The attacker's position and signal strength are set to 30 [m] and 10 [dBm], and all the other conditions are the same. The LiDAR is turned off.

Figure 11 shows the results. The leftmost part shows object detection by the radar, and the central part, footage of the front camera of the ego vehicle, in which the upper half is in the case of the radar alone, and the lower half, the concatenation of the radar and camera. Due to the jamming, it is only when the distance is close to 0 [m] that the radar detects the person in front; It is too late to avoid a crash with the radar alone, while the car is safely stopped with the concatenation thanks to detection by the camera. The comparison proves the effectiveness of the concatenation.

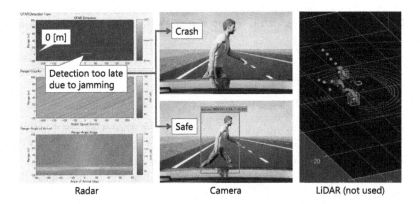

Fig. 11. Evaluation result for detection concatenation. The LiDAR is turned off.

Multi-Object Tracking. As a second example of sensor fusion, we evaluate the multi-object tracking algorithm. Simply put, the algorithm tracks objects by maintaining a list of object detections by multiple sensors. To exclude the effect of misdetections, the algorithm confirms the detection if the same object is detected at least M times out of N sensing periods. Therefore, the greater the ratio M/N is, the more accurate the detection becomes. If we increase N with a fixed ratio M/N, we expect to eliminate the effect of variance and further improve the accuracy, while the algorithm can become more susceptible to attacks due to the increased processing time.

We evaluate the effect of the design parameters M and N in the same attack settings as in Fig. 10. Figure 12 shows the evaluation results for (M, N) = (2, 2) and (9, 12). Overall, (M, N) = (2, 2) is safer than (M, N) = (9, 12) because there are fewer crashes in the former case. However, there are also cases where the brake is applied too soon, which can lead to an uncomfortable driving experience. We can evaluate this kind of trade-off with our simulator.

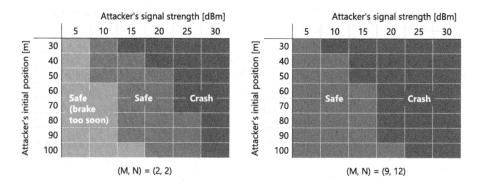

Fig. 12. Evaluation results for multi-object tracking with different sets of parameters

5   Related Work

In terms of sensor attack simulation at system level, the most similar work to our own is Cao et al. [35], which evaluates the impact of sensor attack on LiDAR at the driving decision level. However, the simulator used in their work does not simulate a vehicle dynamics in a physical world unlike the model-based design environment in our work.

Coverage by scenario-based simulation for autonomous systems has been extensively studied [8,9,36–39]. Coverage maximizing techniques include automated generation of test scenarios by random numbers [36,37] and by search algorithms [38]. Abdessalem et al. [38] consider critical test scenarios leading to failures, which looks suitable for performance limitation evaluation. We can also use some prior knowledge about specific systems like AEB to narrow the space of scenarios. Weber et al. [8] define scenarios in a systematic manner with six layers such as road, moving objects and environmental conditions. Our work focuses on the assessment scenarios of AEB in JNCAP and Euro NCAP.

There have been works on safety analysis of autonomous systems using STAMP/STPA [40–42]. While we use conventional testing in this paper, formal methods are also promising for verifying the safety constraints [40,41].

6 Conclusion

We present a framework to evaluate performance limitations of autonomous systems under sensor attacks. Using a prototype simulator of an AEB-equipped car with a radar, a camera and a LiDAR, we show that the framework can identify sensor attack scenarios to be assessed, and evaluate how attacks on the sensors affect the system safety.

Interface between different models and simulators is a key issue in evaluation of autonomous systems, especially when it comes to highly autonomous vehicles becoming more and more complex. We therefore leave it as future work to modularize the sensor attack models, e.g. as FMU (Functional Mock-up Unit), to be used in combination with other simulators. In addition to self-driving cars, there are a diverse range of critical devices and systems that depend on measurement, such as robotic systems, medical devices and control systems. We therefore want to extend our framework to address attacks and countermeasures about measurement interfaces in general: what is called *instrumentation security*.

Acknowledgment. This work is partially based on results obtained from the project (JPNP16007) commissioned by the New Energy and Industrial Technology Development Organization (NEDO).

References

1. Shoukry, Y., Martin, P., Tabuada, P., Srivastava, M.: Non-invasive spoofing attacks for anti-lock braking systems. In: Bertoni, G., Coron, J.-S. (eds.) CHES 2013. LNCS, vol. 8086, pp. 55–72. Springer, Heidelberg (2013). https://doi.org/10.1007/978-3-642-40349-1_4

2. Son, Y., et al.: Rocking drones with intentional sound noise on gyroscopic sensors. In: 24th USENIX Security Symposium (USENIX Security 2015), pp. 881–896. USENIX Association (2015)

3. Chauhan, R.: A platform for false data injection in frequency modulated continuous wave radar all graduate theses and dissertations, 3964 (2014). https://digitalcommons.usu.edu/etd/3964

4. Miura, N., Machida, T., Matsuda, K., Nagata, M., Nashimoto, S., Suzuki, D.: A low-cost replica-based distance-spoofing attack on mmwave FMCW radar. In: Proceedings of the 3rd ACM Workshop on Attacks and Solutions in Hardware Security Workshop (ASHES 2019), pp. 95–100. Association for Computing Machinery (2019)

5. Shin, H., Kim, D., Kwon, Y., Kim, Y.: Illusion and dazzle: adversarial optical channel exploits against lidars for automotive applications. In: Fischer, W., Homma, N. (eds.) CHES 2017. LNCS, vol. 10529, pp. 445–467. Springer, Cham (2017). https:// doi.org/10.1007/978-3-319-66787-4_22

6. Petit, J., Stottelaar, B., Feiri, M., Kargl, F.: Remote attacks on automated vehicles sensors: experiments on camera and lidar. In: Black Hat Europe (2015). https://www.blackhat.com/docs/eu-15/materials/eu-15-Petit-Self-Driving-And-Connected-Cars-Fooling-Sensors-And-Tracking-Drivers-wp1.pdf

7. Liu, J., Yan, C., Wenyuan, X.: Can you trust autonomous vehicles: contactless attacks against sensors of self-driving vehicles. DEF CON (2016). https://doi.org/ 10.5446/36252

8. Weber, H., et al.: A framework for definition of logical scenarios for safety assurance of automated driving. Traff. Injury Prev. **20**(sup1), S65–S70 (2019). PMID:31381437

9. Levermore, T., Peters, A.: Test framework and key challenges for virtual verification of automated vehicles: the VeriCAV project. In: 39th International Conference on Computer Safety, Reliability and Security (SAFECOMP), Position Paper, Lisbon, Portugal, September 2020

10. Kalra, N., Paddock, S.M.: Driving to Safety: How Many Miles of Driving Would It Take to Demonstrate Autonomous Vehicle Reliability? RAND Corporation (2016)

11. ISO: Road vehicles - functional safety. Standard ISO 26262:2018 (2018)

12. ISO: Road vehicles - safety of the intended functionality. Standard ISO/PAS 21448:2019(E) (2019)

13. ISO: Road vehicles - cybersecurity engineering. Standard ISO/SAE DIS 21434:2020(E) (2020)

14. IEC: Fault tree analysis (FTA). Standard IEC 61025:2006 (2006)

15. IEC: Failure modes and effects analysis (FMEA and FMECA). Standard IEC 60812:2018 (2018)

16. Leveson, N., Thomas, J.: STPA handbook. http://psas.scripts.mit.edu/home/get_file.php?name=STPA_handbook.pdf (2018)

17. Tanis, S.: Automotive radar sensors and congested radio spectrum: an urban electronic battlefield? In: Analog Dialogue, vol. 52–57 (2018)

18. Nashimoto, S., Suzuki, D., Miura, N., Machida, T., Matsuda, K., Nagata, M.: Low-cost distance-spoofing attack on FMCW radar and its feasibility study on countermeasure. J. Cryptogr. Eng. (2021)

19. Chen, H.-R.: FMCW radar jamming techniques and analysis (2013). https:// calhoun.nps.edu/handle/10945/37597

20. Thys, S., Ranst, W.V., Goedemé. T.: Fooling automated surveillance cameras: adversarial patches to attack person detection. In: 2019 IEEE/CVF Conference on Computer Vision and Pattern Recognition Workshops (CVPRW), pp. 49–55 (2019). https://www.computer.org/csdl/proceedings/cvprw/2019/1iTvczdcyc0

21. MathWorks: MATLAB. https://jp.mathworks.com/products/matlab.html

22. MathWorks: Simulink. https://jp.mathworks.com/products/simulink.html

23. Epic Games: Unreal Engine. https://www.unrealengine.com/

24. MathWorks: Highway Lane Following. https://jp.mathworks.com/help/mpc/ug/highway-lane-following.html

25. MathWorks: Autonomous Emergency Braking with Sensor Fusion. https://jp.mathworks.com/help/driving/ug/autonomous-emergency-braking-with-sensor-fusion.html

26. MathWorks: Detect and Track Vehicles Using Lidar Data. https://jp.mathworks.com/help/vision/ug/track-vehicles-using-lidar.html

27. MathWorks: Deep Learning with GPU Coder. https://jp.mathworks.com/help/gpucoder/gpucoder-deep-learning.html

28. MathWorks: Radar Signal Simulation and Processing for Automated Driving. https://jp.mathworks.com/help/driving/ug/radar-signal-simulation-and-processing-for-automated-driving.html

29. Bennion, M., Habli, I.: A candid industrial evaluation of formal software verification using model checking. In: Jalote, P., Briand, L.C., van der Hoek, A. (eds.) 36th International Conference on Software Engineering (ICSE 2014), Companion Proceedings, Hyderabad, India, May 31–June 7, 2014, pp. 175–184. ACM (2014)

30. Jeppu, N.Y., Jeppu, Y., Murthy, N.: Arguing formally about flight control laws. In: 2015 International Conference on Industrial Instrumentation and Control (ICIC), pp. 378–383 (2015)

31. MathWorks: Simulink Test. https://jp.mathworks.com/products/simulink-test.html

32. Redmon, J., Farhadi. A.: Yolo9000: better, faster, stronger. In: 2017 IEEE Conference on Computer Vision and Pattern Recognition (CVPR), pp. 6517–6525 (2017)

33. JNCAP. https://www.nasva.go.jp/mamoru/en/

34. Euro NCAP. https://www.euroncap.com/en

35. Cao, Y.: Adversarial sensor attack on lidar-based perception in autonomous driving. In: Proceedings of the 2019 ACM SIGSAC Conference on Computer and Communications Security (CCS 2019), pp. 2267–2281. Association for Computing Machinery (2019)

36. Saigol, Z., Peters, A.: Verifying automated driving systems in simulation: framework and challenges. In: 25th ITS World Congress (2018)

37. Norden, J., O'Kelly, M., Sinha. A.: Efficient black-box assessment of autonomous vehicle safety. CoRR, abs/1912.03618 (2019). http://arxiv.org/abs/1912.03618

38. Abdessalem, R.B., Nejati, S., Briand, L.C., Stifter, T.: Testing vision-based control systems using learnable evolutionary algorithms. In: 2018 IEEE/ACM 40th International Conference on Software Engineering (ICSE), pp. 1016–1026 (2018)

39. Tahir, Z., Alexander, R.: Coverage based testing for v & v and safety assurance of self-driving autonomous vehicles: a systematic literature review. In: 2020 IEEE International Conference on Artificial Intelligence Testing (AITest), pp. 23–30 (2020)

40. Abdulkhaleq, A., Wagner, S., Leveson, N.: A comprehensive safety engineering approach for software-intensive systems based on STPA. In: Proceedings of the 3rd European STAMP Workshop 5–6 October 2015, Amsterdam Procedia Engineering, vol. 128, pp. 2–11 (2015)

41. Dakwat, A.L., Villani, E.: System safety assessment based on STPA and model checking. Saf. Sci. **109**, 130–143 (2018)

42. Ishimatsu, T., Leveson, N.G., Thomas, J., Katahira, M., Miyamoto, Y., Nakao, H,: Modeling and hazard analysis using STPA. In: International Association for the Advancement of Space Safety (IAASS) (2010)

ISO/SAE 21434-Based Risk Assessment of Security Incidents in Automated Road Vehicles

Dominik Püllen$^{(\boxtimes)}$, Jonas Liske, and Stefan Katzenbeisser

University of Passau, Passau, Germany
`dominik.puellen@uni-passau.de`

Abstract. Although numerous automotive security solutions have been presented in the last years, the question of how to properly react to security incidents during vehicle operation has not yet received much attention. In this work, we describe a context-aware scheme for automated road vehicles that assesses the risk of security incidents intending to automatically identify adequate countermeasures. We specifically focus on attack propagation, as related works proved how seemingly uncritical, but compromised vehicle components can cause dangerous situations. Our scheme is inspired by the risk assessment process of the novel ISO/SAE 21434 cybersecurity standard, which uses attack paths to model static threat scenarios. In contrast, our scheme dynamically queries an asset dependency graph once a security incident is reported, in order to identify attack paths leading to pre-assessed damage scenarios. Since the expected damage of a security incident also depends on the vehicle context, we include information such as speed, time, and traffic density into the risk computation. Based on the resulting risk value, the vehicle selects and realizes a compensating (safety) action. Finally, we discuss our scheme and conduct a case study on an automated prototype vehicle.

Keywords: ISO/SAE 21434 · Risk assessment · Context awareness

1 Introduction

The rapidly advancing automation and the increasing number of interfaces have enabled attackers to compromise road vehicles and to take over control. Since then, many tailored defense techniques [4,6,12] have been developed to protect both the passengers and the vehicle. In addition, the automotive industry has begun considering security as an integral component of vehicle engineering [1,10]. Nevertheless, such progress does not imply the absence of security incidents in vehicles, such as an unauthorized access request to safety-critical components or an alarm by an intrusion detection system. Due to a steadily growing attack surface, security incidents will occur, either intentionally or accidentally. For that reason, road vehicles require means to assess and handle them, since not every

© Springer Nature Switzerland AG 2021
I. Habli et al. (Eds.): SAFECOMP 2021, LNCS 12852, pp. 82–97, 2021.
https://doi.org/10.1007/978-3-030-83903-1_6

incident can be ascribed to an attack. For instance, how shall the vehicle behave in case it detects unauthenticated traffic while being in motion? Is an emergency halt necessary if the integrity of the infotainment system cannot be proven? The answer to those questions depends on various factors, such as the vehicle topology, the deployed security concept, and also the vehicle context (e.g., its current speed). A missing answer, however, can lead to unexpected behavior, such as a compromised CD player accelerating the vehicle [19]. Typically, road vehicles consist of many Electronic Control Units (ECUs), whose strong interconnection facilitates the propagation of manipulated data through the vehicle and thus, the propagation of an attack. The most prominent example was given in 2015 by Miller and Valsek [14], who first broke into a poorly secured infotainment system, then manipulated the firmware of a gateway, and eventually gained control of the vehicle by injecting driving commands to the bus. Hence, we argue that the assessment of security incidents has to consider propagating effects, which leads to our main contribution:

We propose a scheme to determine the risk of security incidents during vehicle operation by means of an asset dependency graph. The latter allows identifying attack paths and thus, predict the expected damage through propagation. Our scheme is compliant with the novel ISO/SAE DIS 21434 [10] cybersecurity standard, whose offline risk assessment identifies attack paths to model propagating effects as threat scenarios. While a threat scenario typically describes a static chain of negative actions, we apply the concept of attack path identification to vehicle operation time to assess the risk of security incidents. For a more accurate assessment, we additionally weight the expected damage with vehicle context information. In the end, the computed risk of a security incident leads to a compensating safety action (e.g., driving at reduced speed).

2 Related Work

The automatic analysis of security incidents in computer networks is a well-studied field. Typically, propagation effects are modeled with attack graphs that originate from fault trees, a deductive method to identify the cause of a failure. However, automotive networks have not been the pivot of research, especially when it comes to attack propagation. In the early 2000s, Nikoletseas et al. [15] showed how an intruder with limited power can infiltrate large parts of a computer network through propagation. Noel et al. [16] presented a quantitative analysis of network security risks using attack graphs. They model how various vulnerabilities can be combined for an attack by propagating exploit likelihoods through the attack graph. Roschke et al. [20] deploy attack graphs for an intrusion detection system. They map alerts to graph nodes and detect suspicious graph subsets with a correlation algorithm. Salfer et al. [21] assess the risk of exploits in automotive on-board networks using an automatically generated attack graph. For this purpose, they present a stochastic model that considers both attack vectors and attacker resources more efficiently than Bayesian networks. Krisper et al. [13] align attack events in a graph and compute the cumulative risk with a propagation algorithm. In contrast, we focus on the automotive domain and are compliant with the standardized risk assessment process presented in ISO/SAE 21434.

3   Introduction to ISO/SAE 21434

The upcoming ISO/SAE 21434 is a novel cybersecurity standard for road vehicles that considers the entire vehicle ecosystem. It focuses on cybersecurity relevant items and components inside or on the vehicle perimeter. As the publishing process is still ongoing, we use the approved Draft International Standard (DIS) [10] of ISO/SAE 21434 in the further course of this work, which we refer to as ISO/SAE 21434 for simplicity. ISO/SAE 21434 consists of fifteen clauses that provide common terminology, guidelines for managing security risks, and both cybersecurity policies and processes. A key activity is the risk assessment process of the eighth clause, that typically takes place in the development phase. It enables organizations to identify threat scenarios and to assess their risks, in order to eventually find adequate defense techniques. It is defined as follows: At first, damage scenarios, caused by compromised critical assets, are identified and rated with regard to the expected outcome. The manipulation of such an asset is described by a threat scenario, which in turn consists of at least one attack path, i.e., a chain of dependent actions leading to asset corruption. The combination of the attack path feasibility and the expected damage yields the risk value for the corresponding threat scenario, followed by a risk treatment decision. The latter can involve risk avoidance or reduction, as well as accepting or retaining the risk. The relationship between asset, damage scenario, threat scenario, and attack path is shown in Fig. 1. Eventually, a cybersecurity concept is phrased in the ninth clause by using the methods of clause 8 to assess the risks of items.

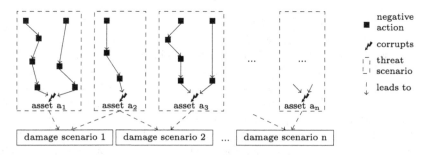

Fig. 1. ISO/SAE 21434 computes the risk of threat scenarios by combining the attack path feasibility with the severity of the expected damage scenarios.

4   Context-Aware Risk Assessments of Security Incidents

In this section, we present our context-aware assessment scheme to handle security incidents in automated vehicles. As mentioned, we follow the risk analysis process of the ISO/SAE 21434 to calculate the risk of security incidents. We consider a security incident as a dynamic threat scenario and use an asset dependency graph to identify and rate possible attack paths. As shown in Fig. 2, our scheme consists of an *offline* and an *online* phase. In the offline phase, damage scenarios, the vehicle topology, and the asset dependency graph are identified and rated. They serve as input for the online phase, which then assesses the risk of a reported security incident and takes a compensating action.

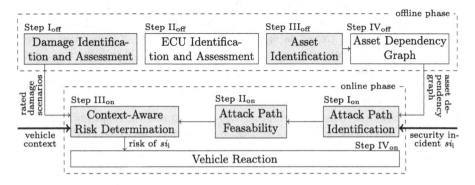

Fig. 2. Structure of our context-aware risk assessment scheme for security incidents in automated road vehicles. White fields are not part of ISO/SAE 21434.

In the remaining sections, we explain each step of our scheme in detail, refer to the corresponding ISO/SAE 21434 requirements, and use the following notation: We denote ECU_m to be the *monitoring* control unit that executes the online phase of our scheme. $\pi_k(x)$ returns the entry k of an ordered collection x (e.g., a tuple).

System Model: Researchers have presented a large number of defense techniques for both legacy and cutting-edge systems [4]. In this work, we expect vehicles to be equipped with state-of-the-art protection techniques that meet the most important automotive security requirements, such as traffic and firmware integrity, timely responses, and access control [18]. We expect all benign ECUs to verify the authenticity of incoming data and discard unverifiable content [6]. On vehicle startup, firmware integrity checks are performed [12] to prevent the unauthorized alteration of ECU firmware [23]. Moreover, we assume that communication delays can be detected by each ECU [22].

Accordingly, we expect that any infringement upon the aforementioned requirements is reported to ECU_m as a security incident. Currently, our work does not cover the case of a manipulated or even dropped security incident. In other words, we assume ECU_m to be always informed about any suspicious and potentially malicious in-vehicle activity. ECU_m is considered a security trust anchor and hence, cannot be corrupted by an attacker. Such anchors typically allow for the isolated execution of code and provide hardware support to maintain the integrity of applications.

Security Incidents: ISO 27005 [9] calls a security incident a negative event that can lead to damage. According to ISO/SAE 21434, we consider a security incident to be a threat scenario initiated by a negative event at vehicle runtime. More precisely, a security incident describes the violation of a protected asset property on an ECU. Such property refers to a threat category, which in turn originates from the deployed threat model. Similar to ISO/SAE 21434, we use the CIA triad, leading to $\mathbb{T} = \{Confidentiality, Integrity, Availability\}$ with $\mathbb{T}$ denoting the set of all threat categories. Upon perception of a negative event, we map it onto a threat category as illustrated in Table 1. Let $\mathbb{E}$ be the set of ECUs and $\mathbb{A}$ be the set of assets. We then formally express a security incident si_i as an element of $\mathbb{E} \times \mathbb{A} \times \mathbb{T}$. That is, si_i consists of the affected ECU, the compromised asset, and the threat category. For instance, the security incident $si_i = (Infotainment, volume\ command, Integrity)$ states that the infotainment system recognized a volume command whose integrity cannot be verified.

Table 1. Negative events are mapped onto a threat category.

Negative event	Threat category (CIA)
Unknown data origin	Integrity
Unverifiable firmware	Integrity
Unauthorized data access	Confidentiality
Delayed traffic	Availability
...	...

4.1 Offline Phase

The offline phase is conducted once by a group of experts, preferably in the vehicle design phase. It provides the necessary input for the online assessment phase, i.e., the rated damage scenarios and the asset dependency graph.

Step I_{off}: Damage Identification and Assessment

At first, we specify the set of all damage scenarios $\mathbb{D}$, which typically happens in an expert panel. A damage scenario describes the expected outcome of a security incident, such as uncontrolled driving behavior. In compliance with

ISO/SAE 21434, each $d_i \in \mathbb{D}$ is ranked in terms of the four SFOP categories Safety, Financial, Operational, and Privacy. More precisely, we use a simple multi-criteria decision-making process that assigns numerical values to each category as explained in [18]. Since we do not find each category equally relevant, we additionally compute distinct weights by implementing a Fuzzy Analytical Hierarchy Process (FAHP) [17]. In that way, a worst-case privacy violation will always be considered less severe than a worst-case safety impact. In the end, the weighted scores are added up, resulting in the four possible values *negligible*, *moderate*, *major*, or *severe*. ISO/SAE 21434 precisely describes the interpretation of each value. For instance, a moderate impact on safety means light injuries, whereas a major impact indicates severe, but non-fatal injuries. At this point, we identify the worst-case damage, which is then later weighted with the vehicle context. In Sect. 5.2 we give examples for this step.

Step II$_{off}$: ECU Identification and Assessment

Next, we specify $\mathbb{E}$ by enumerating all in-vehicle control units. ECUs play a crucial role in our assessment scheme since they do not only trigger damage scenarios but also enable attacks to propagate through the vehicle due to their continuous communication. Therefore, we determine the Attack Potential AP for each ECU$_i \in \mathbb{E}$, which gives us a notion of how likely ECU$_i$ can be manipulated by an attacker. We argue that attacks often require the manipulation of ECUs (e.g., to circumvent security checks), in order to be successfully carried out along a given attack path. AP gives us the possibility for a more realistic computation of the attack path feasibility in Step II$_{on}$. For the determination of AP, we adhere to the Common Methodology for Information Security Evaluation [2], which corresponds to ISO/IEC 18045 [8]. That is, we rank the minimum required attack resources in terms of *Elapsed Time, Expertise, Knowledge, Window of Opportunity*, and *Equipment*. Each parameter is associated with a distinct numerical value as shown in Table 2.

Table 2. CC Methodology to determine the attack potential of an ECU

Elapsed time		Expertise		Knowledge		Windows of opportuniy		Equipment	
Option	Value	Option	Value	Option	Value	Option	Value	Option	Value
<1 week	0	Layman	0	Public	0	Unlimited	0	Standard	0
<1 month	1	Proficient	3	Restricted	3	Easy	1	Specialized	4
<6 months	4	Expert	6	Confidential	7	Moderate	4	Bespoke	7
≤3 years	10	Multiple experts	8	Strictly confidential	11	Difficult	10	Multiple bespoke	9
>3 years	19								

As required by ISO/SAE 18045, we add up those values and obtain an attack score for ECU$_i$, which is subsequently uniformly mapped onto AP as shown in Table 3. In Sect. 5, we conduct this step for ECUs of a prototype vehicle. We suggest doing this step as part of the post-development phase's activities when the ECU's attack potential can be considered fixed.

Table 3. The attack score is transferred to AP, which is used for the computation of attack path feasibility.

Attack score	0–9	10–13	14–19	20–24	>25
Attack feasability	High	High	Medium	Low	Very low
Attack potential AP	0.9	0.7	0.5	0.3	0.1

Step III$_\text{off}$: Asset Identification

In this step, we specify the set of assets $\mathbb{A}$ as required by ISO/SAE 21434. The term asset describes a self-contained unit whose compromise can lead to a damage scenario, either directly or through propagation effects. Thus, assets have to be protected against manipulation, usually by cryptographic and/or anomaly detection systems. The evaluation of popular automotive attacks reveals that attackers typically manipulate in-vehicle traffic, delay communication, and/or intrude ECUs [4,18]. For that reason, we distinguish between *flowing* and *rigid* assets. The former typically concerns logically related traffic between ECUs, such as a service or messages on a specific topic. A rigid asset, in contrast, resides on an ECU, but still can impact out-flowing assets. Most notably, that is the firmware, but cryptographic keys and configuration files may also be considered rigid assets. Furthermore, we distinguish between the two subgroups $\mathcal{A}_t$ and $\mathcal{A}_p$, with $\mathbb{A} = \mathcal{A}_t \cup \mathcal{A}_p$ and $\mathcal{A}_t \cap \mathcal{A}_p = \emptyset$. $\mathcal{A}_t$ contains those assets, whose compromise directly triggers a damage scenario without any detour, i.e., without attack propagation. For instance, the undetected corruption of the steering angle is likely to cause immediate harm. In contrast, $\mathcal{A}_p$ describes assets, whose compromise leads to damage only through attack propagation. For example, a vulnerability in the infotainment firmware may allow an attacker to infiltrate forged commands, but it does not necessarily cause immediate harm. We suggest to iteratively derive $\mathcal{A}_t$ from the previously defined damage scenarios and specify $\mathcal{A}_p$ in an inductive bottom-up approach.

Step IV$_\text{off}$: Asset Dependency Graph

Assets are interdependent because ECUs continuously communicate and perform computations on incoming assets. We write $a_y \leftarrow a_x$ to indicate that any change of a_x also influences a_y. The objective of this step is to arrange the previously identified assets into an Asset Dependency Graph (ADG), which enables us to automatically check whether the corruption of a specific asset can transitively lead to damage. Formally, we describe ADG as a directed multigraph $ADG = (V, E)$. The set of vertices V is made up of physical (V_p) and virtual (V_v) vertices, i.e., $V = V_p \cap V_v$. ECUs are considered physical vertices with in-flowing and out-flowing assets. In contrast, we use a virtual vertex to indicate a dependency on a rigid asset residing on an ECU (e.g., the firmware). An edge $e_i \in E$ exists between a source $\text{ECU}_x \in \mathbb{E}$ and a target $\text{ECU}_y \in \mathbb{E}$ if there is an asset $a \in \mathbb{A}$ flowing from ECU_x to ECU_y. Each edge e_i is associated with a probabilistic weight w, indicating to what extent it contributes to a path in ADG. We express an edge $e_i \in E$ as a quintuplet according to Formula 1.

$$E \subseteq \{(v_x, v_y, a, w, D) \mid (v_x, v_y) \in V^2 \wedge a \in \mathbb{A} \wedge w \in [0, 1] \wedge D \subseteq \mathbb{A} \setminus \{a\}\} \tag{1}$$

An e_i also maintains a reference D containing those assets on which e_i *directly* depends on, i.e., $D = \{a_j \in \mathcal{A}_p \mid a_i \leftarrow a_j \wedge \exists e_j \in E \mid (\pi_a(e_j) = a_j \wedge \pi_{v_y}(e_j) = \pi_{v_x}(e_i))\}$. This is necessary because an out-flowing asset does not necessarily depend on all in-flowing assets. For instance, not every output of an ECU may be secured by a cryptographic key that is stored on that ECU. The specification of ADG requires profound knowledge of the network topology and the in-vehicle data flows, typically known only to the manufacturer. As the creation of such graphs is generally quite labor-intensive, system designers may consider automated approaches [21].

4.2   Online Phase

The online phase is executed by ECU_m every time a security incident is reported. At first, ECU_m identifies attack paths between the corrupted asset and any $a_i \in \mathcal{A}_t$ in ADG. Then, it determines their attack path feasibility and weights the severity of the expected damage with context information, resulting in a risk value of the security incident. Based on this risk, countermeasures are taken.

Step I_{on}: Attack Path Identification

A security incident $si_i \in \mathbb{E} \times \mathbb{A} \times \mathbb{T}$ reports the corrupted asset $a_x = \pi_a(si_i)$ on $ECU_x = \pi_{ECU}(si_i)$. It requires further attention if it can lead to a damage scenario $d_i \in \mathbb{D}$, either directly or through propagation. This is the case if there is a path from ECU_x to an $ECU_y \in \mathbb{E}$, whereas the latter triggers d_i through the corruption of $a_y \in \mathcal{A}_t$, which depends on a_x. We call such a path $pth_{ECU_x-ECU_y}^{a_y \leftarrow a_x}$ and formally describe it as a series of edges in ADG as shown in Eq. 2. For readability reasons we later use pth_i to refer to a valid attack path in ADG.

$$pth_{ECU_x-ECU_y}^{a_y \leftarrow a_x} \in \{(e_1, ..., e_n) \mid e_i \in E\} \tag{2}$$

The boundary conditions of Eq. 2 are given by

$$\pi_{v_x}(e_1) = ECU_x \ \wedge \ \pi_{v_y}(e_n) = ECU_y \ \wedge \ a_y = \pi_a(e_n) \tag{2a}$$

$$\exists e \in E \mid \pi_a(e) = a_x \ \wedge \ \pi_{v_y}(e) = ECU_x \tag{2b}$$

$$\forall e_j, \, _{2 \leq j \leq n} \mid \pi_{v_y}(e_{j-1}) = \pi_{v_x}(e_j) \tag{2c}$$

$$\forall e_j, \, _{2 \leq j \leq n} \mid \pi_a(e_{j-1}) \in \pi_D(e_j) \tag{2d}$$

That is, the path starts at ECU_x and ends at ECU_y, where a_y may be corrupted through attack propagation (2a). Furthermore, there is an edge that leads into ECU_x and contains the corrupted a_x (2b). All edges form a continuous path (2c) between ECUs. Besides, two adjoining edges have to carry dependent assets (2d), since otherwise attack propagation from a_x to a_y would not be possible. Note that ECU_m may find multiple attack paths for a security incident leading to the same damage. In the end, we only consider the most feasible one.

Step II_{on}: Attack Path Feasibility

The ISO/SAE 21434 standard subsequently requires to express the feasibility of each identified attach path as *high, medium, low, very low*. The feasibility gives a notion of the likelihood by which an attack is successfully carried out along a given path. As shown in Eq. 3, we propose to calculate the feasibility $\mathcal{F}_{pth_i}$ of a path $pth_i = (e_1, ..., e_n)$ as the product of the corresponding edge weights w.

$$\mathcal{F}_{pth_i} = F_{\text{map}}\left(\prod_{j=1}^{n} \pi_w(e_j)\right), \text{ with } F_{\text{map}}(p) = \begin{cases} \text{high} & p \in [0.9, 1] \\ \text{medium} & p \in [0.5, 0.9[\\ \text{low} & p \in [0.2, 0.5[\\ \text{very low} & p \in [0, 0.2[\end{cases} \quad (3)$$

The edge weight is only determined in the online phase, since it depends in particular on the security incident and the vehicle state. For instance, a DoS attack is likely to immediately propagate through large parts of the vehicle, as the overall communication slows down. In contrast, illegally injected traffic only becomes harmful if ECUs process it instead of rejecting it. However, according to our system model, the latter only happens if the ECU is compromised. Thus, whenever a path along a specific edge requires the manipulation of an ECU, we take the previously determined attack potential AP from Step II_{off} as w. Besides, edges may be temporarily inactive, especially in a service-oriented environment (e.g., SOME/IP or ASOA [11]). For instance, some services may only run when the vehicle is driving fully automated, while others are put into action for manual maneuvering. Inactive edges are assigned a weight of zero, which makes the attack path infeasible. Altogether, we distinguish between three cases for w:

1. $w = 0$: An edge has a weight of zero, if it is unavailable in the current vehicle state. For instance, a specific service is not running.
2. $w = 1$: An edge weight of one indicates a definite propagation between two vertices in ADG. This happens, for instance, if *Availability* is the threat category of an incident since delayed/dropped messages typically affect all subsequent assets. This also concerns rigid assets, represented by virtual nodes. For example, the output of an ECU always depends on its firmware.
3. $w = \pi_{AP}(\pi_{v_x}(e_i))$: We use the attack potential AP of the edge's source ECU_x as edge weight if the manipulation of ECU_x is required for attack propagation, e.g., to circumvent security checks. This, for instance, may be necessary to transport forged traffic through the vehicle, as benign ECUs would discard it.

Step III_{on}: Context-Aware Risk Determination

Eventually, the risk of the security incident is determined by combining the attack path feasibilities with the expected severity of the associated damage scenarios. Recall that by definition (cf. Step I_{on}) an attack path always leads to exactly one damage scenario. Since we have identified the worst-case damage in Step I_{off}, we inevitably would now obtain a worst-case risk of the security incident, which does not necessarily help in identifying an adequate counter-measure. For instance, the worst-case damage of a corrupted headlight control command is likely to indicate fatalities. This is certainly a good assumption for a night ride and therefore, an emergency stop is plausible. During the day, however, an emergency stop may be excessive, making driving at reduced speed still acceptable (e.g., to head for the next repair shop). To get a realistic notion of the expected damage, we suggest weighting the damage with vehicle context parameters. Note that we are not the first to use such parameters for modeling the vehicle context, although related works typically do this in different fields. For instance, Helmholz et al. [5] consider the daytime and the route frequency for the prediction of trajectories. Since the latter is extraneous for the instant assessment of expected damage, we use the current speed and the traffic density instead. Thus, we express the vehicle context as a vector $C = (\text{S TD T RQ})^{\mathsf{T}}$, consisting of the four parameters *Speed* (S), *Traffic Density* (TD), *Time* (T), and *Route Quality* (RQ). For the sake of simplicity, we only allow two values for each parameter, as shown in Table 4. For example, we distinguish between low speed (<30 km/h) and high speed (≥ 30 km/h), day and night drive, etc. Since those criteria do not all have the same relevance to the severity of a damage scenario, we weight C with the normalized vector $W_{\mathrm{C}} = (0.5\ 0.3\ 0.1\ 0.1)^{\mathsf{T}}$. Thus, the vehicle speed contributes five times more to the context than the route quality does. We obtain $C = C \cdot W_{\mathrm{C}}$, $C \in [0.5, 1]$, a scalar representation of the vehicle context, which we multiply with the numerical damage assessment of $\text{StepI}_{\text{off}}$. In that way, we claim to get a more realistic, context-aware assessment of the expected damage. Finally, we obtain the context-aware risk r_{si_i} for the security incident si_i using the risk matrix in Table 5, proposed by ISO/SAE 21434.

Table 4. Four weighted criteria specify the vehicle context.

	Speed (V)	Traffic Density (TD)	Time (T)	Route Quality (RQ)
Weight	0.5	0.3	0.1	0.1
Value	Low (0.5)	Low (0.5)	Day (0.5)	Easy (0.5)
	High (1)	High (1)	Night (1)	Difficult (1)

Step IV$_{\text{on}}$: Vehicle Reaction

Finally, we map the risk value r_{si_i} to a suitable compensating vehicle action. As previously described, we weight the expected safety implications most compared to financial, operational, privacy consequences. Thus, a safety-related perspective is needed for the reasonable analysis of a security incident. The safety standard ISO 26262, Second Edition, [7] has incorporated considerations of cybersecurity to interact with the ongoing cybersecurity life cycle. Therefore, we claim that there are mechanisms to properly treat safety aspects in a security context. In total, we identified four compensating actions (cf., Table 6) in brainstorming sessions, trying to be consistent with the damage description of ISO/SAE 21434.

Table 5. Risk matrix of ISO/SAE 21434 to determine the risk of a security incident.

r_{si_i}		Attack path feasibility $\mathcal{F}_{pth_i}$			
		Very low	Low	Medium	High
Impact of damage scenario	**Negligible**	1	1	1	1
	Moderate	1	2	2	2
	Major	1	2	3	4
	Severe	1	3	4	5

That means an emergency stop is required if fatalities, severe injuries, and/or large financial losses are expected. We consider driving at reduced speed to be an appropriate measure if operational limitations (e.g., traffic jams) or slight injuries may be the consequence of a security incident. We argue that this option is similar to the run-flat system of contemporary vehicles, which are activated in case of moderate damage. The vehicle displays a dashboard control message if damage to the vehicle and the passenger is only possible with comparably great effort. Additionally, this option is chosen in case of a potential privacy violation. In case of negligible damage, the incident is only logged without additional actions.

Table 6. The risk of the security incidents determines the compensating action.

Risk value	Compensating action
1	Log incident + continue driving without restrictions
2	Log incident + display a dashboard control message
3	Log incident + driving at low speed
4,5	Log incident + emergency stop

5   Case Study and Discussion

In the following, we conduct the *offline* phase for an automated prototype vehicle and discuss the selection of compensating actions by the *online* phase.

5.1   Reference Vehicle

Since 2018, a consortium of seven German universities and selected industry partners have been constructing a fully automated and electric road vehicle with a modularized hardware and software architecture [24]. The objective is to create an evaluation platform for a variety of disruptive automotive concepts from different fields, such as automation, modularization, verification, validation, safety, and security. Since the vehicle is built from scratch, the entire lifecycle can be considered without relying on a legacy system. First and foremost, the vehicle is based on a layered architecture inspired by the human nervous system. That is, the *cerebrum* is responsible for trajectory and behavioral planning. It receives pre-processed radar, lidar, and camera data from four *sensor modules*. Moreover, external servers provide traffic information for the route planning algorithm. The trajectory is then converted to low-level driving commands by the main control system, called *brainstem*. It provides the necessary torques, the rotational frequency, and the steering angle to the *spinal cord* which is comprised of four *dynamic modules* that drive the wheels. Most of the in-vehicle communication relies on the specially developed Automotive Service-Oriented software Architecture (ASOA) [11] which enables flexible communication in Ethernet networks, fast firmware updates, and an easy replacement of hardware components.

5.2   Application of the Offline Phase

We were granted access to a detailed description of the previously described vehicular architecture and the current in-vehicle communication graph. By using the data flows and dependencies between ECUs from this description, we were able to conduct the full offline phase of our scheme. Be aware that we can only present a small subset of our results at this point. All assessments were performed in brainstorming sessions by a group of experts.

As shown in Table 7, we identified 10 damage scenarios and assessed their worst-case impact on the four SFOP categories according to Step I_{off}. Recall that we weight the normalized category scores according to their relevance. The damage scenario d_1 = *Uncontrolled Driving* describes the illegal takeover of control by a malicious party, probably the most dreaded consequence of an attack. In the worst case, not only fatalities are likely to occur, but also severe financial and operational consequences. In contrast, d_3 =*Passenger Inconvenience* describes bothering limitations that mainly originate from the vehicle interior, e.g., a manipulated heating system or blocking doors. Instead of severe physical harm, we rather expect financial damage due to patching and a loss of reputation. According to Step II_{off}, we identified 21 ECUs and determined their probabilistic attack potential AP. For instance, the well protected brainstem yields an absolute attack score of 29, which corresponds to the low attack potential $AP = 0.1$ (cf. Table 3). In contrast, the internal HMI unit has a high attack potential of $AP = 0.9$, mainly due to external interfaces and a steady Internet connection. When it comes to the asset dependency graph, we first specified 50 assets whose corruption can directly cause at least one of the previously identified damage scenarios, i.e., $|\mathcal{A}_t| = 50$. For instance, d_1 = *Uncontrolled Driving* can result from the corruption of the assets a_1 = *Steering Angle*, a_2 = *Torque*, a_3 = *Rotational Frequency*, and/or a_4 = *Firmware Dynamic Module*. After that, we identified 60 assets from $\mathcal{A}_p$ in a bottom-up approach, i.e., those assets that can cause damage through propagation. All in all, our asset dependency graph consists of 173 edges and 59 nodes. Both the fan-in and fan-out of each node

Table 7. In total, 10 damage scenarios have been identified and rated with regard to their worst-case impact on Safety, Financial, Operational, and Privacy.

ID	Damage scenario	S	F	O	P	Weighted score	Result
d_1	Uncontrolled driving	0.5	0.5	0.5	0	0.438	Severe
d_2	Manipulated vehicle routing	0.167	0.167	0.5	0	0.087	Major
d_3	Passenger inconvenience	0.167	0	0.333	0	0.134	Moderate
...	...	...			...		...
d_{10}	Battery degradation	0	0.5	0.333	0	0.078	Major

indicate that most dependencies exist between the main ECUs, i.e., the brainstem, the cerebrum, and the sensor modules. For instance, the brainstem has a fan-out of 26, while smaller ECUs have an average fan-out of only 3. Hence, decent defense techniques are specifically necessary for those ECUs to prevent attack propagation.

5.3   Discussion of the Online Phase

Attack Detection. The online phase assesses the risk of security incidents, which raises the question of which attacks can be unveiled in general. Since a security incident follows a negative event, our scheme can only detect and handle those attacks that can be perceived by the deployed security mechanisms. Thus, a thorough security analysis is already crucial in the vehicle design phase, because only then necessary defense techniques can be identified and later realized. We assign each negative event to a threat category, as this eases the determination and assessment of attack paths. The downside is a reduced attack resolution, as different attacks are mapped on the same threat category. Currently, we deploy the CIA triad and therefore, transfer all security incidents to one of these three categories. The resulting loss of information can be compensated with a more fine-grained threat model, such as STRIDE or the Foundational Requirements of IEC 62443 [3]. However, this inevitability leads to higher model complexity. Moreover, some negative events may not be uniquely related to one threat category. For instance, "unauthorized data access" can affect both *Integrity* and *Confidentiality*. In that case, multiple security incidents could be reported for a single negative event and then be processed by ECU_m according to their risk. In the end, automotive engineers have to decide about the desired trade-off between attack resolution and model complexity.

Time Consumption. Since our scheme assesses security incidents during vehicle runtime, the required time is a crucial parameter, because a delayed compensating action may fail to keep away damage from the passenger. The path identification phase in Step I_{off} traverses the asset dependency graph ADG and stores each *valid* path to an asset $a_i \in \mathcal{A}_t$. As the number of active edges depends on the security incident and the vehicle mode, the graph shape slightly differs for each incident. In practice, either a Breadth First Search or a Depth First

Search allows identifying paths in *ADG*. Assuming *ADG* is internally stored as an adjacency list (i.e., each node keeps a list of adjacent edges), then we need to iterate over the entire list to visit each edge, resulting in a linear complexity of $\mathcal{O}(|V| + |E|)$. While the number of nodes remains constant, the number of *active* edges can change due to the service-oriented communication architecture. Therefore, the runtime of our scheme can be positively affected by keeping the number of edges low, which we will further investigate in future work.

Compensating Actions. Our scheme selects a compensating action based on the computed risk of the security incident. The severity of a security incident depends on the attack path feasibility and the expected damage, whereas the latter additionally takes the vehicle context into account. We simulated security incidents by assuming a corruption of the assets in $\mathcal{A}_t$. Then, we explored how often each compensating action is taken for a fixed worst-case damage (i.e., we assume the most unfavorable vehicle context) but for varying attack path feasibilities. Figure 3 illustrates that smaller feasibility typically leads to a weaker compensating action because more attacker capabilities are required to successfully propagate an attack.

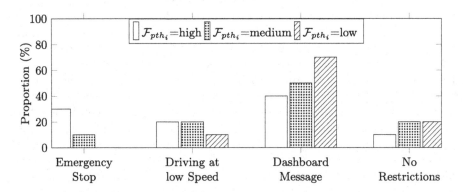

Fig. 3. Assuming a constant worst-case damage, a smaller attack path feasibility usually leads to a weaker compensating action.

More precisely, an emergency stop is necessary for 30% of those security incidents with expected high feasibility. In contrast, no emergency stop is triggered once the feasibility is only low, but instead, weaker actions like dashboard messages are selected (up to 70%). We note that this distribution does not fully reflect reality, as not all assets in $\mathcal{A}_t$ are usually equally often compromised. However, it confirms our intention for an adequate compensating action, i.e., to treat non-critical security incidents less severe than those endangering the passenger's well-being.

Parameter Calibration. A meaningful response to a security incident is only possible with a reasonable calibration of the scheme parameters. For instance, the weights of the SFOP categories or the boundaries between the discrete path feasibility values can be used to tweak the model. We assume that different types of systems and vehicle architectures require different parameter sets. Since we have not yet validated our scheme during vehicle operation time, we justify our parameter selection with a best-practice approach. While the risk matrix is taken from the ISO/SAE 21434 standard, we uniformly map qualitative data to quantitative ones. Moreover, we use FAHP in Step I_{off} to receive weights for the SFOP categories according to their relevance. In that way, we obtain a parameter set that serves as a reasonable starting point and that should be further adjusted during validation.

6 Conclusion

In this work, we presented a context-aware scheme for the risk assessment of security incidents during vehicle operation. The risk is used to automatically select a compensating action to prevent damage while keeping the road vehicle operable as long as possible. We specifically focus on attack propagation, as related works demonstrated how the manipulation of even minor control units may allow an attacker to infiltrate large parts of the vehicle. Our scheme is based on the novel cybersecurity standard ISO/SAE 21434 that uses attack paths to model in-vehicle dependencies for the risk assessment of threat scenarios. This concept is applied to vehicle operation time by treating a security incident as a dynamic threat scenario. Our idea is to identify attack paths within an asset dependency graph and determine their feasibility with respect to the security incident. We applied the offline phase of our scheme to an automated prototype vehicle and manually created an asset dependency graph consisting of 59 nodes and 173 edges. In our case study, we demonstrated that high-risk incidents typically lead to an emergency stop, while less severe incidents trigger weaker actions. We also point out the necessity of effective defense techniques, since they make attack propagation more unlikely and thus, reduce the risk for damage. As a next step, we aim to implement our scheme, validate the parameters, and measure the timing overhead when assessing synthetic incidents.

Acknowledgement. This work has been accomplished within the project "UNICARagil" (FKZ16EMO0392). We acknowledge the financial support for the project by the Federal Ministry of Education and Research of Germany (BMBF).

References

1. SAE J3061 - Cybersecurity Guidebook for Cyber-Physical Vehicle Systems. Standard, Society of Automotive Engineers (2016)
2. Common criteria and common evaluation methodology version 3.1 (2017)
3. ISA-62443 Security for Industrial Automation and Control Systems. Standard, International Society of Automaton (2017)
4. Dibaei, M., et al.: An overview of attacks and defences on intelligent connected vehicles. arXiv preprint arXiv:1907.07455 (2019)

5. Helmholz, P., Ziesmann, E., Robra-Bissantz, S.: Context-awareness in the car: prediction, evaluation and usage of route trajectories. In: vom Brocke, J., Hekkala, R., Ram, S., Rossi, M. (eds.) DESRIST 2013. LNCS, vol. 7939, pp. 412–419. Springer, Heidelberg (2013). https://doi.org/10.1007/978-3-642-38827-9_30
6. Iorio, M., et al.: Securing SOME/IP for in-vehicle service protection. IEEE Trans. Veh. Technol. **69**(11), 13450–13466 (2020)
7. ISO 26262-2:2018: Road vehicles - Functional Safety (2018)
8. ISO/IEC 18045: Methodology for IT security evaluation (2020)
9. ISO/IEC 27005:2018 Information technology-Security techniques - Information security risk management (2018)
10. ISO/SAE DIS 21434:2022(E): Draft international standard ISO/SAE 21434: Road vehicles - cybersecurity engineering (2020)
11. Kampmann, A., et al.: A dynamic service-oriented software architecture for highly automated vehicles. In: 2019 IEEE ITSC, pp. 2101–2108 (2019). https://doi.org/10.1109/ITSC.2019.8916841
12. Kohnhäuser, F., et al.: Ensuring the safe and secure operation of electronic control units in road vehicles. In: IEEE Security and Privacy Workshops, pp. 126–131. IEEE (2019)
13. Krisper, M., Dobaj, J., Macher, G., Schmittner, C.: RISKEE: a risk-tree based method for assessing risk in cyber security. In: Walker, A., O'Connor, R.V., Messnarz, R. (eds.) EuroSPI 2019. CCIS, vol. 1060, pp. 45–56. Springer, Cham (2019). https://doi.org/10.1007/978-3-030-28005-5_4
14. Miller, C., Valasek, C.: Remote exploitation of an unaltered passenger vehicle. Black Hat USA **2015**, 91 (2015)
15. Nikoletseas, S., et al.: Attack propagation in networks. Theory Comput. Syst. **36**(5), 553–574 (2003)
16. Noel, S., et al.: Measuring security risk of networks using attack graphs. Int. J. Next-Gener. Comput. **1**(1), 135–147 (2010)
17. Özdağoğlu, A., Özdağoğlu, G.: Comparison of AHP and fuzzy AHP for the multi-criteria decision making processes with linguistic evaluations (2007)
18. Püllen, D., Anagnostopoulos, N., Arul, T., Katzenbeisser, S.: Safety meets security: using IEC 62443 for a highly automated road vehicle. In: Casimiro, A., Ortmeier, F., Bitsch, F., Ferreira, P. (eds.) SAFECOMP 2020. LNCS, vol. 12234, pp. 325–340. Springer, Cham (2020). https://doi.org/10.1007/978-3-030-54549-9_22
19. Radu, A.-I., Garcia, F.D.: LeiA: a lightweight authentication protocol for CAN. In: Askoxylakis, I., Ioannidis, S., Katsikas, S., Meadows, C. (eds.) ESORICS 2016. LNCS, vol. 9879, pp. 283–300. Springer, Cham (2016). https://doi.org/10.1007/978-3-319-45741-3_15
20. Roschke, S., Cheng, F., Meinel, C.: High-quality attack graph-based ids correlation. Log. J. IGPL **21**(4), 571–591 (2013)
21. Salfer, M., Eckert, C.: Attack graph-based assessment of exploitability risks in automotive on-board networks. In: ARES 2018, pp. 1–10 (2018)
22. Song, H.M., et al.: Intrusion detection system based on the analysis of time intervals of can messages for in-vehicle network. In: 2016 ICOIN, pp. 63–68. IEEE (2016)
23. Ueda, H., Kurachi, R., Takada, H., Mizutani, T., Inoue, M., Horihata, S.: Security authentication system for in-vehicle network. SEI Tech. Rev. **81**, 5–9 (2015)
24. Woopen, T., et al.: UNICARagil-disruptive modular architectures for agile, automated vehicle concepts (2021). https://www.unicaragil.de/en/

Safety and Assurance Cases

Automating the Assembly of Security Assurance Case Fragments

Baoluo Meng$^{(\boxtimes)}$ ⓘ, Saswata Paul ⓘ, Abha Moitra, Kit Siu,
and Michael Durling

General Electric Research, Niskayuna, NY, USA
{baoluo.meng,saswata.paul,moitra,siu,durling}@ge.com

Abstract. This paper presents an approach and tools for automatic generation of security assurance case fragments using patterns for arguing the security of cyber physical systems. The fragments are generated using augmented Goal Structuring Notation (GSN) and can succinctly convey a system's resilience to cyber-threats specified in MITRE's Common Attack Pattern Enumeration and Classification (CAPEC). The GSN schema has been augmented with additional metadata that can be used for visually tracing back to component-level CAPEC threats from higher-level cyber security claims, enabling designers to easily locate flaws in a model when one or more claims cannot be substantiated. An implementation of the approach as a part of the Verification Evidence and Resilient Design in Anticipation of Cybersecurity Threats (VERDICT) toolchain has also been demonstrated along with a case study of a package delivery drone.

Keywords: Security assurance cases · Assurance case patterns · GSN · Security analysis of system architecture · Attack-defense tree · MITRE's CAPEC threats and NIST-800-53 controls

1 Introduction

The failure of safety-critical cyber physical systems can be catastrophic to life, property, or the environment [33]. Therefore, it is imperative that both software and hardware components are subjected to rigorous testing and certification procedures before being approved for use in safety-critical domains such as aviation, medicine, and automotive. However, with increasing complexity of systems, it becomes difficult to accurately and efficiently argue about system guarantees with respect to critical properties. One widely used approach for conveying system guarantees is the construction of *assurance cases* [5]. An assurance case is a structured argument that a system satisfies some desired safety, security, or reliability properties [4]. It is used to convey a compelling and comprehensive case for system guarantees to stakeholders, developers, engineers, and certifiers [20].

There are two main approaches for developing assurance cases – *process based*, which argues that a system has been developed adhering to certain process objectives such as RTCA-DO:178C [30], and *product based*, which argues that a system satisfies certain properties [21]. Several standards such as *Claim, Argument*

© Springer Nature Switzerland AG 2021
I. Habli et al. (Eds.): SAFECOMP 2021, LNCS 12852, pp. 101–114, 2021.
https://doi.org/10.1007/978-3-030-83903-1_7

and Evidence (CAE) [7], *Goal Structuring Notation* (GSN) [23], and *Structured Assurance Case Metamodel* (SACM) [35] exist for representing assurance cases. Traditionally, assurance cases have been used for arguing about the safety of systems. However, there has been recent interest in using them to argue about the security aspects of complex systems [9,29]. Although the fundamental concept of safety and security cases is similar, *i.e.*, using evidences to argue the validity of some claims, according to [4], there are some differences between the two. Security deals with the presence of an intelligent adversary, whereas safety analysis involves assigning probabilities to basic events (e.g., sensor failure), but the same is not possible for adversarial actions; and security-critical systems have to dynamically adapt to adversarial actions.

Model-based development uses tools and techniques to design and analyze domain-specific models of systems [6]. The *Verification Evidence and Resilient Design in Anticipation of Cybersecurity Threats* (VERDICT)[1] is a toolchain for incorporating cyber security resiliency analysis and recommendations in the system design process that are automated, scalable, provide rich feedback, specify trade-offs, and are easy to use by system architects [26,32]. It is available as a plugin for the Open Source AADL Tool Environment (OSATE) [16–18]. The *Model-Based Architectural Analysis* (MBAA) functionality of VERDICT analyzes the security of a system at the architectural level and mainly consists of two tools: STEM (Security Threat Evaluation and Mitigation) [28] and SOTERIA++ [31]. STEM identifies threats from the MITRE Corporation's *Common Attack Pattern Enumeration and Classification* (CAPEC) [1] and selects defenses from the National Institute of Standards and Technology's (NIST-800-53) *Security and Privacy Controls* [2]. SOTERIA++ constructs attack scenarios and highlights areas of the architecture where control measures are either missing or inappropriate, or insufficient rigor has been applied compared to the level of severity of the outcome. It generates attack-defense trees and analyzes them to determine cutsets and the likelihood of successful attacks of top-level events.

The conventional approach for constructing assurance case fragments is informal and manual, making them prone to errors, expensive and time consuming to design, and difficult to manage and evaluate for non-trivial systems [13]. Moreover, the informal nature of arguments is susceptible to logical flaws [5], making them unsuitable for certifying critical applications. This calls for the need of an approach that can holistically combine tools and techniques from formal model-based analysis to automatically generate assurance case fragments, for both safety and security, to aid in system certification. Contributions of this paper towards this end are:

- It augments the traditional GSN schema with metadata to present additional information that can be beneficial for system designers and certifiers.
- It presents security assurance case patterns for automatic instantiation of structured arguments using the augmented GSN schema. The patterns are designed to use domain-specific claims supported by VERDICT and domain-

---

[1] For more details, visit https://github.com/ge-high-assurance/VERDICT.

specific evidences generated by SOTERIA++ for arguing about both security and safety.

- It presents an approach for using the evidences generated by MBAA to construct security case fragments by considering component-level CAPEC threats and NIST-800-53 defenses.
- It demonstrates the effectiveness of the approach by presenting a case study using the model of an unmanned package delivery drone.

The rest of the paper is structured as follows: Sect. 2 describes the augmented GSN schema, security assurance case patterns, and the construction of security case fragments, by using evidences generated by MBAA; Sect. 3 presents a case study using the model of a package delivery drone; Sect. 4 discusses the key findings; Sect. 5 presents related work on assurance cases; and Sect. 6 ends the paper with a conclusion and future directions of work.

2 Contribution

2.1 Augmented Goal Structuring Notation (GSN)

The GSN schema has been widely used in safety case development. It provides a graphical representation of assurance case arguments by using principal elements—*goals, strategies, contexts, assumptions, justifications,* and *solutions,* which are arranged as nodes in a dependency graph. This work augments the traditional GSN to convey additional information to designers and certifiers. We incorporate additional node metadata to provide enhanced visual aid for flaw detection and for generating interactive assurance case fragments with tooltips and clickable links. The metadata also connects the assurance case fragments to the artifacts and evidences generated by VERDICT's MBAA functionality, creating a coherent connection between formal analysis and structured assurance arguments. This augmented GSN schema may also be useful in incorporating defeaters [8] where defeaters capture doubts and objections. The identification of defeaters can surface gaps in an assurance case; and assessment/evaluation of defeaters can strengthen an assurance case.

Traditionally, assurance cases have been designed to substantiate claims using evidences. However, in VERDICT, sometimes the evidences generated may show that a claim cannot be substantiated. Under such circumstances, a traditional GSN fragment cannot be created, but the information about the failure of a claim can still be useful to designers. This information can be used to generate *"incomplete"* GSN fragments which do not argue about the correctness of claims, but provide enhanced visual aid for easily detecting flaws in a model. To this end, we augment the GSN schema by specifying colors for the goal and strategy nodes. Below are the rules determining the color of nodes:

- If a solution fails to substantiate its parent node, then it will be colored red. Otherwise, it is colored green.
- If a goal or strategy node has at least one supporting red node, then it will be colored red. Otherwise, it is colored green.

2.2   Security Assurance Case Patterns

Table 1. A part of the assurance case pattern library for VERDICT.

GSN class	Predefined patterns
Goal	{} "is secure"
	{} "has been mitigated"
Strategy	"Argument: By validity of sub-goals"
	"Argument: By SOTERIA++ analysis of attack-defense trees"
	"Argument: All threats mitigated"
Context	{} "Properties"
	"All applicable threats are identified"
	"Acceptable likelihood =" {}, "Computed likelihood =" {}
	"Acceptable probability =" {}, "Computed probability =" {}
	"A condition and a target probability"
	"A condition and a target likelihood"
Solution	"SOTERIA++ minimum cutset for" {}
	"Evidence that" {} "is secure"

Assurance case patterns [22] are predefined templates that can be instantiated for constructing assurance cases. Patterns restrict the verbiage that can be used for the construction of assurance cases while being flexible enough for expressing structured arguments for a variety of systems. The use of patterns allows the development of consistent assurance case fragments and provides a formal structure for assurance case construction. They are composed of domain-specific constructs that depend on factors such as domain safety and security concerns, type of claims and evidences that can be used, and type of arguments [12].

In order to automatically generate assurance case fragments from the VERDICT toolchain, a library of patterns has been created. These patterns can be used to express structured arguments for claiming cyber-resilience by using evidences generated by SOTERIA++. The patterns allow the implementation to automatically collect information from an *annex*, which is a domain-specific language extension to AADL, and instantiate structured assurance case fragments in the augmented GSN schema. Table 1 shows a part of our assurance case pattern library for the different classes of GSN nodes. The strings inside "" are predefined and the patterns can be used to instantiate the arguments by automatically assigning parameters in place of the curly brackets {}. Using the patterns, it is possible to create claims like "actuation is secure" or "CAPEC-390 has been mitigated" to declare that the actuation component has been secured and that the threat of CAPEC-390 has been mitigated by the implemented defenses respectively. The patterns are specific to the VERDICT toolchain as they can only be used to express statements that make sense in the context of the constructs that are used in VERDICT. *E.g.*, the *verdict* annex supports only two classes of requirements related to security –*mission requirements* and *cyber requirements*–so the patterns can only be used to instantiate requirements belonging to any of these two classes and not arbitrary ones.

2.3   Model-Based Architecture Analysis in VERDICT

One of the fundamental aspects of security, as discussed in Sect. 1, is that it is concerned with external adversarial actions that need to be mitigated. In this section, we will describe the model-based architecture analysis (MBAA) functionality of VERDICT, which can be leveraged to identify cyber vulnerabilities and mitigations, to calculate the likelihood of successful attacks, and ultimately to construct security cases.

Input to Model-Based Architecture Analysis. The input to MBAA is an AADL architecture model annotated with meta-level properties, defense properties, cyber relations and requirements, and mission requirements. Examples of meta-level properties are *componentType* and *connectionType*. Defense properties describe the rigor to implement a defense, which are enumerated type [0; 3; 5; 7; 9]. The implementation rigor is also known as *design assurance level* (DAL), with 0 being the lowest and 9 being the highest rigor. Relations and requirements are written in the *verdict* annex. A mission requirement describes a mission scenario for the system to fulfill, such as delivering a package to an intended location, which is supported by a combination of cyber and safety requirements. The success of mission requirements depends on the success of all supporting cyber and safety requirements. This paper will focus on cyber relations and requirements for security assurance cases. An example of a cyber relation and requirement is illustrated in Fig. 1.

Each cyber requirement is associated with a level of severity (Catastrophic, Hazardous, Major, Minor, and No Effect) corresponding to a quantitative acceptable level of risk ($1e-9$, $1e-7$, $1e-5$, $1e-3$, and $1e-0$, respectively) guided by the standard on Airworthiness Security Methods and Considerations DO-356A.

```
CyberRel "delivery_status_I" = delivery_cmd:I => delivery_status:I;
CyberReq {
    id = "CyberReq01"
    description = "The drone shall be resilient to loss
                   of ability to deliver a package to the
                   appropriate consumer location"
    condition = actuation_out:I or actuation_out:A
                or delivery_status:I or delivery_status:A
    cia = I
    severity = Hazardous
};
```

Fig. 1. A cyber relation and requirement specified in the *verdict* annex.

Note that each exponent of quantitative acceptable level of risk corresponds to a negative DAL value. Each cyber requirement is also tied to a security aspect of the overall system: *Confidentiality (C)*, *Integrity (I)* and *Availability (A)* listed in the *"cia"* field the cyber requirement. The success or failure of the cyber requirement relies on the logical *"condition"*, which describes relevant CIAs of outputs of the system, and will be used in attack-defense tree analysis in SOTERIA++. Cyber relations describe how various possible threats associated with confidentiality, integrity, or availability propagate through components. The cyber relation in Fig. 1 tells that any threats affecting the integrity of the input *delivery_cmd* of the component will also affect integrity of the output *delivery_status*.

Analyzing the Security of System Architectures. The analysis of the security of system architectures replies on two components of VERDICT: Security Threat Evaluation and Mitigation (STEM) and SOTERIA++, which are described below.

Security Threat Evaluation and Mitigation (STEM). Security Threat Evaluation and Mitigation (STEM) [28] is a SADL-based semantic model with a set of rules to identify vulnerabilities and suggest defenses. SADL (Semantic Application Design Language) [10] is an English-like language based on Web Ontology Language (OWL) for building semantic models and authoring rules. STEM takes system architectural information annotated with properties as input, identifies possible threats from MITRE's CAPEC, and suggests mitigations from NIST-800-53 Security and Privacy Controls. There are a total of 61 Meta Attack Pattern CAPECs, but not all of them are relevant to embedded systems of interest to STEM. STEM incorporates 37 Meta Attack Pattern CAPECs that are relevant to an embedded system. Mitigations are linked to CAPECs so that controls are only suggested if they are useful in mitigating attacks that have a defined effect on the system under consideration. STEM encompasses three types of rules to identify CAPEC vulnerabilities, suggest NIST mitigations, and associate NIST mitigations with defense properties for attack scenarios. An example of a STEM rule to identify CAPEC-131 is shown in Fig. 2.

Rule **Vul-CAPEC-131-1**
if **oneOf(componentType** of a **Subsystem,**
  Software, SwHwHybrid, SwHumanHybrid, Hybrid)
and **insideTrustedBoundary** of a **Subsystem** is true
and **dataReceivedIsUntrusted** of the **Subsystem** is true
then **applicableCM** of the **Subsystem** is CAPEC-131A-ResourceAvailability.

Fig. 2. An example of a STEM rule to identify CAPEC-131 Resource Leak Exposure.

The SOTERIA++ Tool. Once STEM identifies all *applicable* CAPEC attacks and *implemented* NIST controls for components or connections of an architecture model, the information is fed into SOTERIA++ [31] for further analysis. The *severity* field of a cyber requirement defines the top-level cyber security goal of a system, which corresponds to a quantitative acceptable level of risk. The acceptable level of risk indicates that the likelihood of successful attack should be less than or equal to one failure for every 10^d hours of system operation, where d is a DAL value. SOTERIA++ analyses the system to determine whether it satisfies the top-level goal or not. Starting with the *condition* field of a cyber requirement, it back-traces each atomic *output:CIA* expression via cyber relations and connections to construct an attack-defense tree. The non-leaf node of an attack-defense tree is a logical (AND, OR, or NOT) operator; the leaf node is either an attack or a mitigated attack. Note that the NOT operator can only be applied to defenses, and is used for convenience to convert the DAL of defense to a likelihood (10^{-DAL}). An attack-defense tree is used to compute the likelihood of a successful attack against a system. It represents possible ways to violate a cyber requirement for a system in which an adversary is attempting attacks given the implementation rigors of defenses. This is the reason we compute likelihoods and not probabilities. Worse case is assumed for attacks. As a result, the likelihood of attack is always 1. This then leads the designer to focus on applying rigor to the defense implementation in order to lower the likelihood of an attack. The implementation is a defense profile, which is a conjunction or disjunction set of defenses. The likelihood of a conjunction set of defenses (i.e., all defenses are required to mitigate an attack) is the maximum of the defense DALs; the likelihood of a disjunction set (i.e., any defense in the profile by itself can mitigate an attack) is the minimum. The intuition behind the min/max calculation is to encourage the designer to focus on the most rigorous defense. The likelihood of an attack is calculated by doing a minimum of the attack (which is always 1) and the likelihood of the set of defenses. An attack-defense tree is said to be mitigated if the calculated likelihood of defenses is less than or equal to the severity level assigned for the top-level goal for a system; thus the cyber requirement corresponding to the attack-defense tree is satisfied.

2.4 Security Assurance Case Construction

To construct meaningful and accurate security assurance case fragments, it is important to ensure that there is a succinct argument that shows how a cyber requirement is supported by the mitigations suggested by STEM. However, a cyber requirement can be dependent upon several components and each component can be vulnerable to multiple CAPEC threats. Embedding all the assurance information for a cyber requirement into a single security case fragment may involve a GSN with hundred of nodes, making the fragment difficult to comprehend and visualize in physical or even digital form. Therefore, a security case fragment for every cyber requirement is decomposed into two classes of fragments:

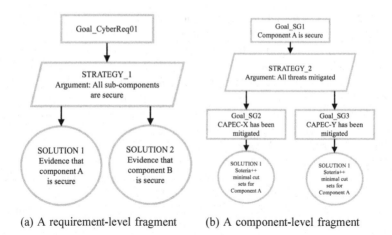

(a) A requirement-level fragment   (b) A component-level fragment

Fig. 3. Sample structures of a requirement-level and a component-level security case fragment.

- *Requirement-level fragments* (Fig. 3a) - There is a single requirement-level fragment for a cyber requirement whose root goal claims that the requirement has been satisfied. The solution nodes provide evidences for the security of every component that affects the requirement. Clicking on a solution node causes associated component-level fragments to be displayed. The root goal is supported by a strategy that argues that all sub-components are secure.
- *Component-level fragments* (Fig. 3b) - Each solution node for a requirement-level fragment is a separate component-level fragment whose root goal claims that the component is secure. The root goal has a sub-goal for every CAPEC threat that the component is vulnerable to, each of which claims that the threat has been mitigated. The sub-goals are directly supported by solution nodes which are evidences generated by SOTERIA++. The root goal uses a strategy which argues that all threats have been mitigated. The solution node for a CAPEC threat shows the computed and acceptable likelihoods for the threat, and the implemented NIST defenses when hovered upon.

3   Case Study: A Delivery Drone

To demonstrate the effectiveness of our approach, we use the AADL model of an unmanned package delivery drone. The architecture of the delivery drone is provided in Fig. 4 and its AADL model is publicly available on Github (see footnote 1). The drone has been designed to perform tasks that have certain associated mission and cyber requirements. In VERDICT, cyber requirements are independent of each other while mission requirements are dependent on a combination of cyber requirements. The requirements for the delivery drone have been provided by security domain experts and have been specified in the *verdict* annex of the AADL model of the drone.

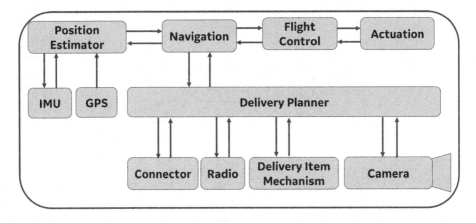

Fig. 4. The delivery drone system architecture.

Consider cyber security requirement CyberReq01 in Fig. 1, which states that the *"The drone shall be resilient to the loss of ability to deliver a package to the appropriate consumer location."* To fulfill this requirement, it has been determined by a security expert that the integrity and availability of *actuation_out* of *actuation* and *delivery_status* of *delivery item mechanism* component shall not be compromised. Moreover, the consequence of failure is hazardous, which corresponds to an acceptable likelihood of failure 1e−07. To declare a system to be secure in the context of VERDICT, we need to show that all applicable CAPECs to its subcomponents have been sufficiently mitigated by the implemented defenses in the system. Specifically, all the subcomponents propagating threats that could affect the integrity and availability of those outputs have been secured. The AADL model annotated with properties is then fed into the VERDICT tool for analysis. A requirement-level security case fragment for CyberReq01 is returned as shown in Fig. 5.

It is evident from the figure that CAPEC threats to the *actuation* component are not all mitigated. When clicking on the solution node for the *actuation* component, a component-level fragment is displayed as Fig. 6, which shows all applicable CAPEC threats to the component and that the specific CAPECs have not been mitigated. In this case, actuation is susceptible to CAPEC-390 (Bypassing Physical Security), which has not been sufficiently mitigated, as the computed likelihood for a successful CAPEC-390 attack is 1e−05, which is greater than the acceptable likelihood 1e−07. In other words, the defense (NIST-SE-3 System Access Control) has only been implemented to the design assurance level of 5, which is not sufficient to mitigate CAPEC-390 to avoid hazardous consequences. The minimal custset evidence computed by SOTERIA++ can be viewed by hovering over or clicking on the solution node. In practice, what happens at this point is that the security expert works with the system design team to upgrade the defense. Upgrading the defense means revisiting the list of required design activities and performing additional activities to gain credit for a higher

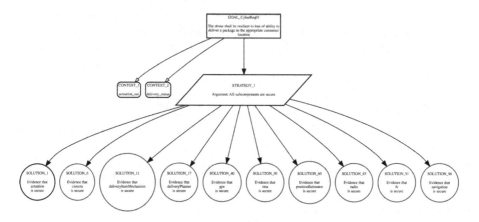

Fig. 5. The requirement-level security fragment for CyberReq01.

assurance level. Activities include traditional software assurance items such as configuration management, requirements management, requirements verification, and testing. The list also includes specific security assurance activities such as penetration testing and vulnerability assessment. These activities are incorporated into the System Engineering design process of the drone. The system designer can then update the defenses from design assurance level 5 to 7, and then run the new model again with VERDICT to show that the system is secure.

4   Discussion

Security cases allow developers to easily detect security flaws in the design of security-critical systems, thereby enabling them to go back to the design phase and include security features that were initially missed. They also allow capturing the rationale behind adding more stringent security features that give a clear benefit and those for which the benefit is not very obvious [4]. While developing the capability to generate security cases from the VERDICT toolchain, we considered the best ways to enable end-users to easily detect and isolate security issues in the design of a system. One of the unique aspects of the security case generation feature of the VERDICT toolchain is the ability to trace back to the source of potential security issues by visual inspection of the colored branches of the GSN fragments. This not only saves time but also makes it easy for users who are not experienced with the intricacies of the toolchain to understand the cause of a problem and provide specific feedback to address the same.

Developing security cases is fairly complicated as they have to consider the high uncertainties involved with how an intelligent adversary may attack a system. Therefore, they should be designed in a way so as to indicate the cause of possible security flaws with as much specificity as possible. This is because sometimes, adding a feature to avoid one security flaw may lead to the creation of another flaw that was originally absent. Therefore, an accurate error-tracing

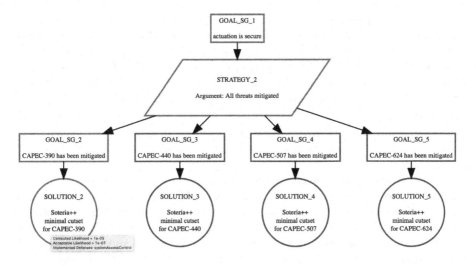

Fig. 6. The component-level security fragment for the *actuation* component.

capability is an important factor in security cases. Automated security case patterns should be designed in such a way so that they can provide enough information to capture the essence of an issue. Modularity is important as it allows the security case fragments to be inspected and reviewed independently, reducing the cognitive workload on the users. This also allows the cases to be aggregated into larger fragments as necessary when trying to analyze a system from varying levels of granularity. Another important aspect to keep in mind while designing automatic security cases is to limit the amount of additional information that developers have to include in the specification. This helps to keep the specifications concise and free from additional content that is not critical to the design.

5   Related Work

Denney *et al.* [12–15] present AdvoCATE, a toolset for automatic creation and management of safety cases. Resolute [19] is an assurance case language and tool where users manually formulate claims and the rules for justifying them. Meng *et al.* [25] formalize queryable safety assurance case in SADL language and use evidence from fault tree analysis to substantiate claims. Assurance cases have traditionally been used for arguing about system safety. Existing work on security cases has primarily investigated the efficacy of using assurance cases for arguing about security. Alexander *et al.* [4] provide a detailed analysis of the benefits of using assurance cases for arguing about system security concerns. They point out the fundamental differences between safety and security concerns that make it necessary to consider external adversarial threats in security cases. Agudo *et al.* [3] investigate the development of security cases by mapping different stages of the system development life-cycle to the structure of the security cases.

Vivas *et al.* [34] present a similar approach for integrating security cases with the security engineering phase of development in which the argument structure of a security claim represents the structure of the system development phase. Poreddy and Corns [29] argue the security of a generic avionics mission controls system by analyzing the potential threats to the mission control computer and constructing arguments for tangible claims in the GSN form. The AMASS Tool Platform [11] provides a novel holistic approach to support assurance and certification for Cyber Physical Systems. Yamamoto and Kobayashi [24] propose the creation of security assurance cases for arguing the security of mobile architecture models.

We improve upon existing work by creating domain-specific patterns to automatically generate security case fragments using MITRE's CAPEC threats and NIST's cyber defenses. The toolchain implementation also provides advantages such as automatic pattern instantiation, automatic generation of assurance case fragments, user-defined argument instantiation, integration and management of formal method tools based evidences, modularity of fragments, and easy management of fragments. XML artifacts are generated for the fragments that can be consumed by other tools for assimilation of lower-level fragments to create higher-level fragments, archiving, and further ontological and formal analysis.

6 Conclusion and Future Work

This paper presented an approach for constructing security case fragments using an augmented GSN schema that can present useful information to designers even if one or more claims cannot be substantiated. The fragments can be automatically generated using assurance case patterns without requiring any manual formulation or specification from the users. The security case fragments consider the component-level CAPEC vulnerabilities of a system to argue about the cyber requirement level claims. An implementation of the approach was presented as a functionality of the VERDICT toolchain and a case study was presented to show the effectiveness of the approach. A potential application of the tool would be to assemble security cases to certify systems towards cyber-security standards such as ASTM-F3286 Standard Guide for Cybersecurity and Cyberattack Mitigation.

One potential future direction would be to extend STEM to include more threats and defenses that target at other areas such as network communications, or integrate with tools like CyVAF [27] that can cover a broader range of threats and defenses. Also, we would like to investigate how the evidence generated by other formal methods tools can be used for constructing security assurance case fragments. This would involve expanding the current assurance case pattern library to accommodate the verbiage of other tools.

Acknowledgement. Distribution Statement "A" (Approved for Public Release, Distribution Unlimited). This research was funded by the Defense Advanced Research Projects Agency (DARPA). The views, opinions and/or findings expressed are those of the authors and should not be interpreted as representing the official views or policies of the Department of Defense or the U.S. Government.

References

1. Common Attack Pattern Enumeration and Classification (CAPEC) (2017). https://capec.mitre.org
2. Security and Privacy Controls for Information Systems and Organizations (2017)
3. Agudo, I., Vivas, J.L., López, J.: Security assurance during the software development cycle. In: Proceedings of the International Conference on Computer Systems and Technologies and Workshop for PhD Students in Computing, pp. 1–6 (2009)
4. Alexander, R., Hawkins, R., Kelly, T.: Security Assurance Cases: Motivation and the State of the Art. The University of York, York (2011)
5. Bagheri, H., Kang, E., Mansoor, N.: Synthesis of assurance cases for software certification. In: Proceedings of the International Conference on Software Engineering (2020)
6. Basir, N., Denney, E., Fischer, B.: Deriving safety cases for hierarchical structure in model-based development. In: Schoitsch, E. (ed.) SAFECOMP 2010. LNCS, vol. 6351, pp. 68–81. Springer, Heidelberg (2010). https://doi.org/10.1007/978-3-642-15651-9_6
7. Bloomfield, R., Netkachova, K.: Building blocks for assurance cases. In: 2014 IEEE International Symposium on Software Reliability Engineering Workshops, pp. 186–191. IEEE (2014)
8. Bloomfield, R., Rushby, J.: Assurance 2.0: A manifesto (2020)
9. Cheah, M., Shaikh, S.A., Bryans, J., Wooderson, P.: Building an automotive security assurance case using systematic security evaluations. Comput. Secur. **77**, 360–379 (2018)
10. Crapo, A., Moitra, A.: Toward a unified English-like representation of semantic models, data, and graph patterns for subject matter experts. Int. J. Semant. Comput. **7**(03), 215–236 (2013)
11. De La Vara, J., Parra, E., Ruiz, A., Gallina, B.: The amass tool platform: an innovative solution for assurance and certification of cyber-physical systems. In: Joint 26th International Conference on Requirements Engineering: Foundation for Software Quality Workshops, Pisa, Italy, vol. 2584. CEUR-WS (2020)
12. Denney, E., Pai, G.: A methodology for the development of assurance arguments for unmanned aircraft systems. In: 33rd International System Safety Conference (ISSC 2015) (2015)
13. Denney, E., Pai, G.: Automating the assembly of aviation safety cases. IEEE Trans. Reliab. **63**(4), 830–849 (2014)
14. Denney, E., Pai, G.: Tool support for assurance case development. Autom. Softw. Eng. **25**(3), 435–499 (2017). https://doi.org/10.1007/s10515-017-0230-5
15. Denney, E., Pai, G., Pohl, J.: AdvoCATE: an assurance case automation toolset. In: Ortmeier, F., Daniel, P. (eds.) SAFECOMP 2012. LNCS, vol. 7613, pp. 8–21. Springer, Heidelberg (2012). https://doi.org/10.1007/978-3-642-33675-1_2
16. Feiler, P.: The Open Source AADL Tool Environment (OSATE). Technical report, Carnegie Mellon University (2019)
17. Feiler, P.H., Gluch, D.P.: Model-Based Engineering with AADL: An Introduction to the SAE Architecture Analysis & Design Language. Addison-Wesley, Boston (2012)
18. Feiler, P.H., Gluch, D.P., Hudak, J.J.: The architecture analysis & design language (AADL): An introduction. Technical report, Carnegie Mellon University (2006)
19. Gacek, A., Backes, J., Cofer, D., Slind, K., Whalen, M.: Resolute: an assurance case language for architecture models. ACM SIGAda Ada Lett. **34**(3), 19–28 (2014)

20. Graydon, P.J.: Formal assurance arguments: a solution in search of a problem? In: 2015 45th Annual IEEE/IFIP International Conference on Dependable Systems and Networks, pp. 517–528. IEEE (2015)
21. Guerra, S., Sheridan, D.: Compliance with standards or claim-based justification? The interplay and complementarity of the approaches for nuclear software-based systems. In: Proceedings of the Twenty-Second Safety-Critical Systems Symposium, Brighton, UK (2014)
22. Hawkins, R., Clegg, K., Alexander, R., Kelly, T.: Using a software safety argument pattern catalogue: two case studies. In: Flammini, F., Bologna, S., Vittorini, V. (eds.) SAFECOMP 2011. LNCS, vol. 6894, pp. 185–198. Springer, Heidelberg (2011). https://doi.org/10.1007/978-3-642-24270-0_14
23. Kelly, T., Weaver, R.: The goal structuring notation-a safety argument notation. In: Proceedings of the Dependable Systems and Networks 2004 Workshop on Assurance Cases, p. 6. Citeseer (2004)
24. Kobayashi, N., Morisaki, S., Yamamoto, S.: Mobile security assurance for automotive software through ArchiMate. In: You, I., Leu, F.-Y., Chen, H.-C., Kotenko, I. (eds.) MobiSec 2016. CCIS, vol. 797, pp. 10–20. Springer, Singapore (2018). https://doi.org/10.1007/978-981-10-7850-7_2
25. Meng, B., et al.: Towards developing formalized assurance cases. In: 2020 AIAA/IEEE 39th Digital Avionics Systems Conference (DASC), pp. 1–9 (2020). https://doi.org/10.1109/DASC50938.2020.9256740
26. Meng, B., et al.: VERDICT: a language and framework for engineering cyber resilient and safe system. Syst. **9**(1) (2021). https://doi.org/10.3390/systems9010018. https://www.mdpi.com/2079-8954/9/1/18
27. Meng, B., Smith, W., Durling, M.: Security threat modeling and automated analysis for system design. SAE Int. J. Transp. Cyber Privacy **4** (2021). https://doi.org/10.4271/11-04-01-0001
28. Moitra, A., Prince, D., Siu, K., Durling, M., Herencia-Zapana, H.: Threat identification and defense control selection for embedded systems. SAE Int. J. Transp. Cyber. Privacy **3** (2020)
29. Poreddy, B.R., Corns, S.: Arguing security of generic avionic mission control computer system (MCC) using assurance cases. Proc. Comput. Sci. **6**, 499–504 (2011)
30. RTCA-DO: 178c: Software considerations in airborne systems and equipment certification (2011)
31. Siu, K., Herencia-Zapana, H., Prince, D., Moitra, A.: A model-based framework for analyzing the security of system architectures. In: 2020 Annual Reliability and Maintainability Symposium (RAMS), pp. 1–6. IEEE (2020)
32. Siu, K., et al.: Architectural and behavioral analysis for cyber security. In: 2019 IEEE/AIAA 38th Digital Avionics Systems Conference (DASC), pp. 1–10. IEEE (2019)
33. Sommerville, I.: Software Engineering (2011). ISBN-10 137035152, 18
34. Vivas, J.L., Agudo, I., López, J.: A methodology for security assurance-driven system development. Requir. Eng. **16**(1), 55–73 (2011)
35. Wei, R., Kelly, T.P., Dai, X., Zhao, S., Hawkins, R.: Model based system assurance using the structured assurance case metamodel. J. Syst. Softw. **154**, 211–233 (2019)

Safety Case Maintenance: A Systematic Literature Review

Carmen Cârlan[1]([✉]), Barbara Gallina[2], and Liana Soima[1]

[1] fortiss GmbH, Munich, Germany
{carlan,soima}@fortiss.org
[2] Mälardalen University, Västeras, Sweden
barbara.gallina@mdh.se

Abstract. Safety standards from different domains recommend the execution of a process for keeping the system safety case up to date, whenever the system undergoes a change, however, without providing any more specific guidelines on how to do this. Even if several (semi)automated safety case maintenance approaches have been proposed in the literature, currently, in the industry, the execution of this process is still manual, being error prone and expensive. To this end, we present in this paper the results of what is, to the best of our knowledge, the first Systematic Literature Review (SLR) conducted with the goal to provide a holistic overview of state-of-the-art safety case maintenance approaches. For each identified approach, we analyze its strengths and weaknesses. We observe that existing approaches are pessimistic, identifying a larger number of safety case elements as impacted by a change than the number of the actually impacted elements. Also, there is limited quantitative impact assessment. Further, existing approaches only address a few system change scenarios when providing guidelines for updating the safety case.

Keywords: Safety case maintenance · Systematic literature review

1 Introduction

Motivation: The system safety case can be used as a medium for assessing the impact system changes have on the system safety assurance [40]. Safety cases are explicitly required or recommended by standards from different safety critical domains, such as ISO 26262 voluntary standard [39] and UL 4600 [41] for automotive systems, the CENELEC EN 50129 standard for railway systems, the IAEA's safety standards for systems based on nuclear energy, the FAA Order 8900.1 FSIMS, Vol. 16, in the avionics or the JSP 318B standard for military aircraft systems. A safety case is a specialization of an assurance case, which is an argumentation that, based on certain evidence, a system satisfies certain system requirements, in a defined operational environment [6]. As a small change to any related safety work product may affect a large part of the safety case [25,41], the same standards also require that the system safety case reflects the current

© Springer Nature Switzerland AG 2021
I. Habli et al. (Eds.): SAFECOMP 2021, LNCS 12852, pp. 115–129, 2021.
https://doi.org/10.1007/978-3-030-83903-1_8

status of the system. For example, ISO 26262 states that the safety case is a work product (e.g., hazards list, requirements specification, system design) generated by the execution of the system safety lifecycle and that the evidence in the safety case is a compilation of the other safety work products. The same standard, in Part 10, recommends maintaining the system safety case consistent with the other safety work products. According to ISO 26262 and to UL 4600, safety case maintenance is a two-phased process. First, given a change in a safety work product, a change impact analysis (CIA) shall be conducted. Second, based on identified impact, the safety case shall be updated correspondingly. However, even if the maintenance of safety cases is a complex process, neither ISO 26262, nor any standard in other domains provide guidelines on concrete techniques for executing safety case maintenance. Currently, in practice, safety case maintenance is manually executed by safety engineers, being an error-prone and time and resource consuming process. Consequently, the inadequate management of changes in the specification of the system or its operational context has led in the past to accidents [4] or NHTSA recalls[1]. Automated change impact analysis for the system safety case and the existence of guidelines for how to update it given certain types of system changes would be beneficial. As such, safety case maintenance approaches have gained much attention in research.

Objectives and Method: The scope of this work is (1) to synthesize a comprehensive list of all automated safety case maintenance approaches proposed in the literature in the time interval 2000–2020, based on the results of a conducted systematic literature review (SLR) [26]; and (2) to report on the results of an in-depth analysis of these approaches. We are especially interested in assessing the following capabilities: 1) the degree of automation for CIA, 2) the accuracy of CIA, 3) the provision of support for quantitative CIA, 4) the provision of guidance for updating the safety case, 5) the availability of tool support. Further, we also analyze the addressed change scenarios.

Results: The SLR resulted in the selection of 65 papers, presenting 26 approaches for safety case maintenance. The results of our conducted SLR highlight three important limitations of existing approaches for safety case maintenance. First, we conclude that existing approaches are pessimistic, as their identified impact area of impact may be larger than the actual impact area. Second, we observe that, in current literature, guidelines for updating the safety case only address a few change scenarios. Another outcome of our analysis is that there is limited support for assessing the impact of a change in a quantitative manner.

The remainder of this paper is organized as follows. In Sect. 2, we provide essential background on safety case maintenance and an overview of related SLRs and mapping studies. In Sect. 3, we describe the protocol we used for this SLR. Then, in Sect. 4, we synthesize and analyze existing work regarding maintaining the system safety case consistent with other system and safety engineering artifacts in the literature. Towards the end, in Sect. 5, we summarize the results of our SLR, while highlighting the limitations of current approaches, and, in Sect. 6, we conclude by proposing possible research directions.

---

[1] https://betterembsw.blogspot.com/p/potentially-deadly-automotive-software.html

2 Background and Related Work

Safety Case Maintenance - Basic Concepts. Kelly and McDermid [25] elaborate on the two–phased safety case maintenance process recommended by ISO 26262. They classify *impacted* elements in two categories: elements *directly impacted* by a system change (also called *challenged* safety case elements) and *indirectly impacted* elements, which are only impacted due to the "ripple effect" of the impact propagation. Kelly and McDermid also differentiate between CIAs that only identify *potentially* impacted elements, which still need to be manually checked by the safety engineer, and accurate CIA that only identify *actually* impacted elements, which are surely invalidated by the change. Potentially impacted elements may be either actually impacted elements or false positives, meaning that they might not actually be invalidated by the change.

Related Literature Studies. While there are two works reporting on the state of the art in safety case tools [29], and in safety case languages [21], to the best of our knowledge, there is no review of current safety case maintenance approaches. Maksimov et al. [29] report the results of a survey concerning tool support for the creation and management of safety cases, while also analyzing certain tool functionalities, among which the support provided for maintaining safety case models consistent with other work products. Govardhanrao [21], in her master thesis, presents the results of a comparative analysis scoping a selection of argumentation languages. Among others, she assesses the support for consistency checks between system design and developing safety cases and the support offered to automatically update the safety argumentation, given system changes. The two related reviews have certain important limitations. On the one hand, the list of tools identified by Maksimov et al. is outdated – five relevant tools have been reported in the literature after the publication of this survey, namely in 2019 and 2020. On the other hand, the analysis performed by Govardhanrao does not cover all the existing approaches, but a selection of those. Further, none of the two works differentiates between the capability of identifying directly impacted elements, and the capability of automatically computing impact propagation, nor between accurate and inaccurate change impact analyses. Finally, Maksimov et al. and Govardhanrao do not provide details regarding the system change scenarios that are regarded by the maintenance approach.

3 Review Protocol

3.1 Establishing the Quasi-Gold Standards by Manual Search

We first selected several publications as our quasi-gold standards (QGSs) [45]. We base our SLR on the results of the SLR conducted by Maksimov et al. [29]. As such, to establish our QGSs, we first started by manually selecting publications, which, according to Maksimov et al., present tools that have medium or strong support for safety case maintenance [29]. While Maksimov et al. identify 17 publications describing 17 different tools implementing approaches for safety case maintenance, after applying our filtering criteria (see Subsect. 3.3), we only selected 13 from these publications. Based on our expert knowledge, to these 13 selected publications we also added 3 publications, each discussing another approach different to the ones identified by Maksimov et al. [29] and 2 deliverables presenting different capabilities of the AMASS platform [14].

We complemented the aforementioned QGSs with publications we manually selected. To this end, we conducted a manual search through the proceedings published in the time frame 2000–2020 of a selection of venues that we identified as highly relevant for the safety engineering research: International Conference on Dependable Systems and Networks (DSN), International Conference on Reliable Software Technologies (Ada-Europe), European Dependable Computing Conference (EDCC), International Conference on Computer Safety, Reliability and Security (SAFECOMP), Pacific Rim International Symposium on Dependable Computing (PRDC), The International Conference/Workshop on High-Assurance Systems Engineering (HASE), International Symposium on Software Reliability Engineering (ISSRE), International Symposium on Model-Based Safety and Assessment (IMBSA), International Conference on Software Engineering (ICSE), International Conference on Model Driven Engineering Languages and Systems (MODELS) and their satellite workshops. After filtering out based on the criteria presented in Subsect. 3.3, we identified 19 publications presenting additional capabilities of already identified approaches, but also introducing 6 new approaches.

3.2   Automated Search and Snowballing

Next, while using a search string, we automatically search through the following databases: ACM Digital library, IEEE, Springer, Elsevier, Google Scholar and dblp. Based on the most frequent words found in the publications we included in our QGS, we specified the following search string: ("safety case" AND "maintenance") OR ("assurance case" AND "maintenance") OR ("safety case" AND "change") OR ("assurance case" AND "change") OR ("safety case" AND "evolution") OR ("assurance cases" AND "evolution"). The search resulted in 3 selected publications. To mitigate the potential limitations due to blurry terminology, we then applied snowballing [44], which resulted in the selection of 10 more relevant publications describing approaches already identified during the other search phases.

3.3   Exclusion and Inclusion Criteria

During our search, we only selected publications whose title and abstract made it explicit that the publication was presenting the results of primary research on approaches for safety case maintenance, or at least for safety case change impact analysis. Further, we also used some other inclusion criteria. First, we only regarded the publications that appeared in 2000 or after. We chose 2000 to be the earliest date for our search, since this was the earliest publication year of one paper which we identified as QGS. We searched all publications that appeared until December 2020 - as the automatic and manual searches were finished in January 2021. Second, given a number of different papers presenting the same approach, we considered all the papers, in order to ensure that we do not miss any information regarding safety case maintenance support. We excluded publications matching any of the following criteria: 1) publications presenting approaches for assurance case modeling, but not having at least a minimum support for assurance case maintenance; 2) publications describing maintenance approaches for other types of assurance cases (e.g., security or trust cases), 3) publications presenting support for safety case maintenance only as future work; 4) books, tutorials or poster publications; 5) publications that have not been peer-reviewed; 6) publications that are only available in the form of abstracts/posters and presentations, 7) publications not written in English.

3.4   Evaluation Criteria

After identifying all existing approaches for safety case maintenance, we carried out an in-depth analysis of these approaches, while using a set of evaluation criteria (see Table 1-a). In conformance with the work of Kelly and McDermid [25], we differentiate between approaches that support the automated identification of challenges, i.e., of the safety case elements directly impacted by a system change (**EC1**) and the automated identification of indirectly impacted elements due to impact propagation (**EC2**). **EC3** addresses the accuracy of CIA (i.e., freedom of false positives). Inspired by one work from our QGS, namely the one of Jaradat and Bate [24], who propose a quantitative assessment of the change impact, we defined evaluation criterion **EC4**. In accordance to the requirements of ISO 26262 and UL4600, evaluation criterion **EC5** assesses the capability of approaches to provide guidance for updating the safety case in accordance to the CIA results. Further, as UL4600 highly recommends the usage of tools to execute impact analysis, we are also interested whether the identified approaches have tool support (**EC6**).

According to ISO 26262 and UL 4600, given a change in the system specification, the safety case needs to be re-evaluated. Different system change scenarios have a different impact on the system safety case [1]. As such, we analyze the change scenarios addressed by the identified safety case maintenance approaches, especially considering the scenarios in Table 1-b. **CS1-CS3** are general change scenarios. However, since addressing more concrete change scenarios increases the accuracy of CIA [27], we also address more concrete scenarios. A report on an industrial survey conducted by de la Vara et al. [13] presents the state of the practice with respect to safety evidence change impact analysis. The survey reports that requirement specifications are the artifacts most exposed to changes during the entire system lifecycle (**CS4**). UL4600 requires that safety case maintenance is executed given any change in the system design (**CS5**). ISO 26262 mandates the demonstration that all the safety critical requirements specified for the system under consideration have been designed, implemented and tested. This is usually established by traceability links. The report of de la Vara et al. emphasizes the fact that traceability links are bound to frequently undergo changes during the entire system lifecycle. This is because changes in different engineering artifacts also trigger changes in the traceability links (**CS6**). The results of the industrial survey conducted by de la Vara et al. also indicate that safety analysis is frequently re-executed during the system lifecycle, outputing new analysis results (**CS7**). Another type of engineering artifacts reported by the survey conducted by de la Vara et al. as undergoing frequent changes are verification and validation results (**CS8**). Whenever the system requirements, design or source code change, re-verification shall be executed, in order to detect early specification violations. Further, UL4600 requires the execution of safety case maintenance given changes to the intended operational environment (i.e., contextual assumptions) (**CS9**). UL4600 also recommends that the impact a system reconfiguration has on the system safety case needs to analyzed (**CS10**).

Table 1. Overview of the used evaluation criteria and the addressed change scenarios.

a) Evaluation Criteria		b) Addressed change scenarios	
ID	**Evalation Criteria**	**ID**	**Change Scenario**
EC1	Support for automated challenge detection	CS1	Deletion of any system artifact
EC2	Support for automated impact propagation	CS2	Addition of any system artifact
EC3	Accuracy of CIA (i.e., freedom of false positives)	CS3	Modification of any system artifact
		CS4	Modification of a requirement
		CS5	Modification of the system architecture
EC4	Support for quantitative impact assessment	CS6	Modification of traceability links within system artifacts
EC5	Support for updating the safety case	CS7	Modification of risk assessment (i.e., ASIL assignments, according to ISO 26262)
EC6	Tool support	CS8	Addition of new verification evidence
		CS9	Modification of operational environment
		CS10	Modification of system parameter values

4 Review Results

Our SLR resulted in 65 selected publications, presenting 26 different approaches for safety case maintenance. We present the identified approaches in Table 2, together with an overview of their capabilities for keeping the system safety case consistent with other safety artifacts. While for the analysis of each approach we used all the publications we found during our literature search, in the table we only reference one or two most relevant publications, due to space restrictions.

Commercial Approaches. Our SLR revealed the existence of two safety case maintenance approaches implemented in commercial tools. Both approaches support the traceability between safety cases and other safety artifacts. NOR-STA [43] addresses change scenario **CS1**, by identifying missing traceability links. Further, NOR-STA recommends, given the addition of system model elements, the addition of argumentation legs concerning the newly added model

Table 2. Overview of the identified safety case maintenance approaches.

Approach	Automated challenge detection	Automated impact propagation	False positives	Quantitative CIA	Update guidance	Tool support
AC-ROS [10]	Yes [CS3]	No	No	No	Yes	Yes
AdvoCATE [15]	No (safety case regeneration) [CS2]	–	No	No	No	Yes
AF3 [8]	Yes [CS1, CS3]	Yes	Yes	No	No	Yes
AMASS Platform [14, 20, 22]	Yes [CS1, CS2, CS3, CS4]	No	Yes	No	No	Yes
ASCE [17, 32]	Yes [CS1, CS3]	No	Yes	No	No	Yes
Checkable Safety Cases [9]	Yes [CS1, CS2, CS5, CS6, CS7]	Yes	No	No	Yes	Yes
D-CASE [18, 31]	Yes [CS3, CS4, CS10]	No	Yes	No	Yes	Yes
DMILS [11]	No (safety case regeneration) [CS2]	–	No	No	No	Yes
Dynamic Safety Cases [2]	Yes [CS3]	Yes	No	Yes	Yes	No
ENTRUST [7]	Yes [CS3, CS4, CS5]	No	Yes	No	No	Yes
ETB [12]	Yes [CS4, CS5]	No	Yes	No	Yes	Yes
Event-B Extension [35]	Yes [CS4, CS5]	No	Yes	No	Yes	Yes
GAGE [5]	Yes [CS5]	No	Yes	No	Yes	Yes
HIP-HOPS extension [36]	Yes [CS4, CS5]	No	Yes	No	Yes	Yes
Interlocking Safety Cases [42]	Yes [CS3]	No	No	No	Yes	Yes
Isabelle/SACM [33]	Yes [CS1, CS2, CS3, CS4, CS5]	No	Yes	No	Yes	Yes
MMINT-A [27, 37]	Yes [CS1, CS3, CS7]	Yes	Yes	No	No	Yes
NOR-STA [43]	Yes [CS1, CS2, CS3]	No	No	No	Yes	Yes
Resolute [19]	Yes [CS5]	No	Yes	No	Yes	Yes
SAFA [1]	Yes [CS1, CS2, CS3, CS6]	No	Yes	No	Yes	Yes
Safety Cases for IMS [34]	Yes [CS2, CS3]	Yes	Yes	No	Yes	No
Safety Case Synthesis [3]	Yes [CS4, CS5]	No	Yes	No	Yes	Yes
SAM [25]	Yes [CS1, CS2, CS3]	Yes	Yes	No	Yes	Yes
SANESAM [24]	Yes [CS5]	Yes	No	Yes	No	Yes
SPIRIT [28]	Yes [CS1, CS2 CS3, CS6]	Yes	Yes	No	No	Yes
Weaving Safety Cases [23]	No (safety case regeneration) [CS2, CS6]	–	No	No	No	Yes

element (**CS2**). The Assurance and Safety Case Environment (ASCE) [17, 32] reflects the impact of modifications in referenced files on the safety case model. Further, ASCE supports the comparison between two structured safety cases, by specifying each version of a safety case as a Kripke structure.

Change Impact Propagation in Safety Cases. There are several approaches supporting change impact propagation, i.e., which identify the safety case elements indirectly impacted by a change. The approach of Nicholson et al. [34] is only presented in an abstract manner, and is only adequate for Integrated Modular Systems (IMS). The Sensitivity ANalysis for Enabling Safety Argument Maintenance (SANESAM) [24] is an accurate safety case maintenance approach specifying any system change as the modification of the failure rates of hardware components. SANESAM is a quantitative CIA and also provides support for impact propagation. AutoFOCUS3 (AF3) [8], SAM [25], SPIRIT [28] and Model Management INTeractive for Assurance cases (MMINT-A) [27] offer support for automated identification of challenged safety case elements and automated change impact propagation given the deletion (**CS1**) or modification of any referenced artifact (**CS3**). However, all four approaches are prone to output false positives. Additionally, MMINT-A checks for correct Automotive Safety Integrity Level (ASIL) decomposition, given changes in the ASIL attribute of safety case goals (**CS7**). Checkable Safety Cases [9] is a novel approach for accurately checking the consistency between safety case and other system models, supporting different change scenarios (**CS1-CS5**).

Automated (Re)generation of Safety Cases via Formal Methods. Several works, such as ETB [12], GAGE [5], the extension of Event-B for safety case modeling [35], SACM/Isabelle [33], Resolute [19], the safety case synthesis approach proposed by Bagheri et al. [3], and the HIP-HOPS extension for modeling safety cases [36] propose the usage of formal verification methods for the specification of entire safety cases, and/or the formal specification of system safety properties referenced in the safety case. Some of these approaches even support the automated generation of the system safety case, by instantiating patterns with information from formal verification engines. The satisfaction of formally specified safety claims can be verified against a certain system specification (model or code) and the obtained verification results can be automatically integrated as evidence in the system safety case. Given a change in a formally specified safety requirement (**CS4**), it is checked if the system architecture or the code (still) implements the respective requirement. However, the impact of that change on the rest of the artifacts (e.g., hazards lists) is not assessed. Further, given a change in the system architecture (**CS5**), some of these approaches identify the impacted safety case claims and suggest for reverification. However, given counter-evidence or additional evidence (**CS8**), which has not been referenced in the safety case before, there is no support for change impact propagation throughout the rest of the argumentation. All existing approaches may output false positives, as not every system change invalidates the verification evidence.

(Re)generation of Safety Cases via Automated Pattern Instantiation. Approaches such as the weaving safety case models approach proposed by Hawkins [23], DMILS [11], and AdvoCATE [15] remove the need for change impact analysis altogether by instead regenerating the impacted part of the assurance case model, based on automated pattern instantiation (see Table 2). The automated pattern instantiation is done by the usage of a third model (i.e., an instantiation model) containing the mappings between pattern parameters to be instantiated and the values with which they shall be instantiated. On the one hand, in the approach proposed by Hawkins [23], the parameters may be instantiated with direct traceability links to other safety engineering models. Therefore, given the deletion (**CS1**) or the modification (**CS3**) of a referenced system model, the impact of the change on the safety case model is automatically reflected and the patterns are automatically re-instantiated. However, the approach is too pessimistic, triggering the need for re-instantiation whenever a system change occurs, even when the change does not impact the validity of the safety argumentation. On the other hand, in DMILS and AdvoCATE the instantiation models only contain an ID or name of the referenced model elements, instead of having a direct traceability link. Therefore, the user of AdvoCATE needs to manually assess the impact of a system change on the safety case and decide if the patterns shall be re-instantiated. However, AdvoCATE supports the automated identification of outdated verification evidence and automated integration of regenerated verification results as evidence in the system safety case. None of these two approaches provides guidance for how to update the assurance case given invalidated claims or evidence.

Safety Cases Updated at Runtime. Some approaches such dynamic safety cases proposed by Denney et al. [16], interlocking safety cases [42], ENTRUST [7], AC-ROS [10] and D-CASE [30] support the automated update of assurance cases at runtime, based on the feedback received from online monitors. However, these frameworks only address changes of certain system configuration parameters at runtime (specialization of **CS3**) and do not provide any solution for handling changes of other assurance artifacts such as hazards, or requirements. Only the dynamic safety cases approach supports to some extent change impact propagation, either by computing how the confidence level is affected by a parameter change or by propagation based on the relations between GSN elements.

Change Impact Analysis for Safety Cases. The Architecture-driven, Multi-concern and Seamless Assurance and Certification of Cyber-Physical Systems (AMASS) platform [14] and the Safety Artifact Forest Analysis (SAFA) [1] approach are unique approaches, which cannot be fit only into one of the categories above. AMASS supports the identification of invalidated verification evidence due to changes in system specification [22] (change scenarios **CS4, CS5**) and change impact analysis given changes in the features of the systems [20]. AMASS provides some support for updating the safety case by updating the contracts of a system component, given changes in the contextual assumptions [38]. SAFA automatically generates GSN structures based on a model specifying the traceability links among different safety artifacts. Further, SAFA can compare two different GSN structures in order to support the assessment of the evolution of a safety case by identifying the elements added (**CS2**), deleted (**CS1**) or modified (**CS3**) in a new version of the same GSN structure. SAFA also provide guidance for updating the safety case, given certain change scenarios, namely **CS1, CS2, CS3, CS4**. The two approaches are bound to output false positives, and do not support automated computation of impact propagation.

Threats to Validity. One of the main threats to validity is that our results may be incomplete. To address this threat, we used a hybrid search strategy, combining manual and automated searches with snowballing. However, our results may be unreliable due to our lacunary interpretation of the capabilities of some of the approaches, especially of the ones for which little information was available. *Internal validity.* We conducted our SLR based on a defined review protocol, as recommended by Kitchenham and Charters [26]. Further, the selection of relevant publications was peer reviewed. While the third author of this paper executed all the search phases presented before, after the execution of each search phase, the first author of this paper reviewed all the exclusions and the final set of included papers. Towards the end of the SLR, the second author of this paper checked, agreed upon, and refined the whole set of extracted data. *External validity.* Threats to validity such as bias in data selection, extraction, and classification may impair the generalizability of the results. While we aimed at providing complete and valid results, the SLR protocol presented in Sect. 3 could be used for further updates and/or replication reviews to reinforce its results. For example, the review could be complemented by searching for approaches for the maintenance of assurance cases addressing other types of requirements, such as security, dependability, or trustworthiness.

5   Discussion

Inaccurate Automated CIA. The results of our analysis show that all identified approaches have some support for automated detection of change impact, by exploiting traceability links between safety case elements and other engineering artifacts. Only the approaches for (re)generating safety cases via automated pattern instantiation remove the need for change impact analysis altogether. However, 15 out of 26 identified approaches only support the identification of challenged safety case elements, namely the ones directly impacted by a system change, without also computing the impact propagation throughout the entire safety argumentation. Further, with few exceptions, most of the approaches are inaccurate, namely they are prone to output false positives (see Fig. 1-a). The approaches that do not provide false negatives only focus on very specific types of changes or are only adequate to be used for specific types of systems.

Lack of Support for Quantitative CIA. Only 2 out of 26 approaches provide support for the quantitative assessment of change impact (see Fig. 1-b). The dynamic safety cases approach proposes the assessment of the impact a system change has on the confidence in the safety argumentation. However, there are no details provided on the implementation of this assessment. SANESAM computes the impact of system changes on the results of a failure probability analysis. However, such analysis can only be done for hardware components.

Limited Support for Updating the Safety Case. In Fig. 1-c, we see that only for certain system change scenarios some guidance for updating the impacted safety case is provided, whereas 9 approaches do not offer any guidance. 7 out of 26 approaches propose re-verification given changes in either the system requirements or in the implementing system specification (i.e., system architecture, system configuration, or source code). However, these approaches do not give any guidelines on what to do if the newly generated verification results are negative or if additional evidence (i.e., evidence that has not been referenced in the safety case) is generated. Further, these approaches are bound to output false positives, meaning that they cannot determine if the verification evidence actually needs to be re-generated, given a certain change. Another type of safety case update recommendation is provided by SAFE, MMINT-A and NOR-STA. Given the addition of a new element in a referenced set of elements, these approaches can identify that the argumentation is incomplete, and suggest the addition of new claims in the argumentation regarding the newly added elements. Further, all the approaches for maintaining safety cases consistent with the system configuration at runtime propose to switch from one safety case to another, in correspondence to the system re-configuration.

Few Addressed Change Scenarios. Each of the state-of-the-art approaches for safety case maintenance addresses one or more change scenarios (see Fig. 1-f). However, not every change scenario we identified as relevant for current practice in Table 1 is addressed by current approaches. Some change scenarios are poorly addressed in the literature. While approaches such as ETB, GAGE, Event-B Extension, Resolute, HIP-HOPS extension, ENTRUST and SAFA can detect

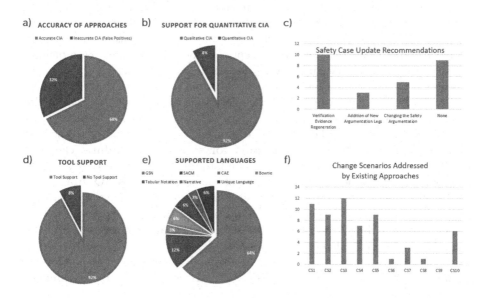

Fig. 1. Statistics on existing safety case maintenance approaches.

the addition of new verification evidence, which they integrate in the safety case (**CS8**), they do not assess the extent of the impact the new evidence has on the safety argumentation. Moreover, to our knowledge, there is no approach addressing the modification of contextual assumptions (**CS9**). Currently, given the modification of a contextual assumption, the entire argumentations depending on that assumption needs to be manually checked by the safety engineer.

Tool Support. Most of the identified approaches have some tool support (see Fig. 1-d). The approach proposed by Nicholson et al. [34] and the dynamic safety cases proposed by Denney et al. do not provide tool support, and also their usage is not exemplified, leaving certain open questions regarding how to actually apply them. **Used safety case languages.** 19 of the identified approaches can be applied for safety case models compliant with the GSN (see Fig. 1-e).

6   Summary and Future Lines of Work

In the recent years, due to the stringent practical needs for automating safety assurance, we have witnessed a boom in state-of-the-art approaches for automated safety case maintenance. In this paper, we reported on the results of a systematic literature review, which we conducted to identify all the existing approaches for safety case maintenance. These results may be extended by also searching for maintenance approaches for any type of assurance cases. Further, another possible extension of the SLR is to also consider product-line oriented approaches, which integrate maintenance of safety cases with variability approaches. The SRL resulted in the selection of 65 papers, presenting 26 different approaches for safety case maintenance, within the interval 2000–2020. While analyzing the strengths and weaknesses of the identified approaches, we identified a set of literature gaps, from which we drew some future lines of work. **More Accurate CIAs.** With few exceptions, the existing safety case maintenance approaches are inaccurate. More accurate CIA, requiring less involvement of safety engineers, would be beneficial as it would decrease the time and effort needed for the execution of the whole safety case maintenance process. Quite recently, it was estimated that, the change of one line of code in an avionics system costs around 1 million dollars, and that it takes approximately one year to be implemented[2]. **Increased support for safety case update.** Our SLR results showed that guidance for how to update the safety case is only available for few change scenarios, which may have serious consequences[3]. **Quantitative CIAs.** There are few approaches assessing the impact of a change in a quantitative manner. Quantitative analyses could provide the safety engineers with a better understanding of the implications of a certain change on the system safety, especially in the context of systems dominated by uncertainty [2]. **Addressing more change scenarios.** According to our analysis results, there is a lack of support in handling change scenarios **CS5-CS10** specified in Table 1. Similar

---

[2] https://insights.securecodewarrior.com/one-line-of-code-1-million/.

[3] https://libertyvillepersonalinjurylawyer.com/software-fault-liability/.

to Kokaly et al. [27], we believe that these gaps could be covered if safety case maintenance approaches would focus on more concrete change scenarios, and by enhancing safety case models with metadata specifying the sensitivity of the safety case to specific system changes.

References

1. Agrawal, A., Khoshmanesh, S., Vierhauser, M., Rahimi, M., Cleland-Huang, J., Lutz, R.R.: Leveraging artifact trees to evolve and reuse safety cases. In: Proceedings of the 41st International Conference on Software Engineering, pp. 1222–1233. IEEE/ACM (2019)
2. Asaadi, E., Denney, E., Menzies, J., Pai, G.J., Petroff, D.: Dynamic assurance cases: a pathway to trusted autonomy. Computer **53**(12), 35–46 (2020)
3. Bagheri, H., Kang, E., Mansoor, N.: Synthesis of assurance cases for software certification. In: Proceedings of the 42nd International Conference on Software Engineering, New Ideas and Emerging Results, pp. 61–64. ACM (2020)
4. Betz, J., Heilmeier, A., Wischnewski, A., Stahl, T., Lienkamp, M.: Autonomous driving-a crash explained in detail. Appl. Sci. **9**(23), 13–20 (2019)
5. Björnander, S., Land, R., Graydon, P., Lundqvist, K., Conmy, P.: A method to formally evaluate safety case arguments against a system architecture model. In: Proceedings of the 2nd Edition of the Workshop on Software Certification. IEEE Computer Society (2012)
6. Bloomfield, R.E., Bishop, P.G.: Safety and assurance cases: past, present and possible future - an adelard perspective. In: Dale, C., Anderson, T. (eds.) Making Systems Safer - Proceedings of the 18th Safety-Critical Systems Symposium, pp. 51–67. Springer, London (2010). https://doi.org/10.1007/978-1-84996-086-1_4
7. Calinescu, R., Weyns, D., Gerasimou, S., Iftikhar, M.U., Habli, I., Kelly, T.: Engineering trustworthy self-adaptive software with dynamic assurance cases. IEEE Trans. Softw. Eng. **44**(11), 1039–1069 (2018)
8. Cârlan, C., Nigam, V., Voss, S., Tsalidis, A.: Explicitcase: tool-support for creating and maintaining assurance arguments integrated with system models. In: Proceedings of the 38th International Symposium on Software Reliability Engineering Workshops, pp. 330–337. IEEE (2019)
9. Cârlan, C., Petrisor, D., Gallina, B., Schoenhaar, H.: Checkable safety cases: Enabling automated consistency checks between safety work products. In: Proceedings of the 31st International Symposium on Software Reliability Engineering - ISSRE Workshops, pp. 295–302. IEEE (2020)
10. Cheng, B.H.C., Clark, R.J., Fleck, J.E., Langford, M.A., McKinley, P.K.: AC-ROS: assurance case driven adaptation for the robot operating system. In: Proceedings of the 23rd International Conference on Model Driven Engineering Languages and Systems, pp. 102–113. ACM (2020)
11. Cimatti, A., DeLong, R., Marcantonio, D., Tonetta, S.: Combining MILS with contract-based design for safety and security requirements. In: Koornneef, F., van Gulijk, C. (eds.) Combining MILS with contract-based design for safety and security requirements. LNCS, vol. 9338, pp. 264–276. Springer, Cham (2015). https://doi.org/10.1007/978-3-319-24249-1_23
12. Cruanes, S., Hamon, G., Owre, S., Shankar, N.: Tool integration with the evidential tool bus. In: Giacobbazzi, R., Berdine, J., Mastroeni, I. (eds.) VMCAI 2013. LNCS, vol. 7737, pp. 275–294. Springer, Heidelberg (2013). https://doi.org/10.1007/978-3-642-35873-9_18

13. de la Vara, J.L., Borg, M., Wnuk, K., Moonen, L.: An industrial survey of safety evidence change impact analysis practice. IEEE Trans. Software Eng. **42**(12), 1095–1117 (2016)

14. De La Vara, J.L., Parra, E., Ruiz, A., Gallina, B.: The AMASS tool platform: an innovative solution for assurance and certification of cyber-physical systems. In: CEUR Workshop Proceedings, vol. 2584. CEUR-WS (2020)

15. Denney, E., Pai, G.: Tool support for assurance case development. Autom. Softw. Eng. **25**(3), 435–499 (2018)

16. Denney, E., Pai, G.J., Habli, I.: Dynamic safety cases for through-life safety assurance. In: Proceedings of the 37th International Conference on Software Engineering, pp. 587–590. IEEE Computer Society (2015)

17. Felici, M.: Modeling safety case evolution – examples from the air traffic management domain. In: Guelfi, N., Savidis, A. (eds.) RISE 2005. LNCS, vol. 3943, pp. 81–96. Springer, Heidelberg (2006). https://doi.org/10.1007/11751113_7

18. Fujita, H., Matsuno, Y., Hanawa, T., Sato, M., Kato, S., Ishikawa, Y.: DS-bench toolset: tools for dependability benchmarking with simulation and assurance. In: Proceedings of the 42nd International Conference on Dependable Systems and Networks, pp. 1–8. IEEE Computer Society (2012)

19. Gacek, A., Backes, J., Cofer, D.D., Slind, K., Whalen, M.: Resolute: an assurance case language for architecture models. Computing Research Repository (CoRR), abs/1409.4629 (2014)

20. Gallina, B.: AMASS Deliverable: design of the AMASS tools and methods for cross/intra-domain reuse. Technical report D6.3, AMASS Consortium (2018)

21. Govardhanrao, S.B.: A comparative analysis of argumentation languages in the context of safety case development. Master's thesis, Mälardalen University, School of Innovation, Design and Engineering (2019)

22. Grüber, T.: AMASS Deliverable: Prototype for multi-concern assurance. Technical Report D4.6, AMASS Consortium (2018)

23. Hawkins, R., Habli, I., Kolovos, D., Paige, R., Kelly, T.: Weaving an assurance case from design: a model-based approach. In: Proceedings of the 16th International Symposium on High Assurance Systems Engineering, pp. 110–117. IEEE (2015)

24. Jaradat, O.T.S., Bate, I.: Using safety contracts to guide the maintenance of systems and safety cases. In: Proceedings of the 13th European Dependable Computing Conference, pp. 95–102. IEEE Computer Society (2017)

25. Kelly, T.P., McDermid, J.A.: A systematic approach to safety case maintenance. Reliab. Eng. Syst. Safety **71**(3), 271–284 (2001)

26. Kitchenham, B., Charters, S.: Guidelines for performing systematic literature reviews in software engineering (EBSE 2007-001) (2007)

27. Kokaly, S., Salay, R., Chechik, M., Lawford, M., Maibaum, T.: Safety case impact assessment in automotive software systems: an improved model-based approach. In: Tonetta, S., Schoitsch, E., Bitsch, F. (eds.) SAFECOMP 2017. LNCS, vol. 10488, pp. 69–85. Springer, Cham (2017). https://doi.org/10.1007/978-3-319-66266-4_5

28. Lin, C.-L., Shen, W., Yue, T., Li, G.: Automatic support of the generation and maintenance of assurance cases. In: Feng, X., Müller-Olm, M., Yang, Z. (eds.) SETTA 2018. LNCS, vol. 10998, pp. 11–28. Springer, Cham (2018). https://doi.org/10.1007/978-3-319-99933-3_2

29. Maksimov, M., Fung, N.L.S., Kokaly, S., Chechik, M.: Two decades of assurance case tools: a survey. In: Gallina, B., Skavhaug, A., Schoitsch, E., Bitsch, F. (eds.) SAFECOMP 2018. LNCS, vol. 11094, pp. 49–59. Springer, Cham (2018). https://doi.org/10.1007/978-3-319-99229-7_6

30. Matsuno, Y.: A design and implementation of an assurance case language. In: Proceedings of the 44th Annual International Conference on Dependable Systems and Networks, pp. 630–641. IEEE Computer Society (2014)
31. Matsuno, Y., Yamamoto, S.: A framework for dependability consensus building and in-operation assurance. J. Wirel. Mobile Netw. Ubiquit. Comput. Depend. Appl. **4**(1), 118–134 (2013)
32. Mistry, M., Felici, M.: Implementation of change management in safety cases. Formal Aspects of Safety-Critical Systems (2008)
33. Nemouchi, Y., Foster, S., Gleirscher, M., Kelly, T.: Mechanised assurance cases with integrated formal methods in Isabelle. Computing Research Repository (CoRR), abs/1905.06192 (2019)
34. Nicholson, M., Conmy, P., Bate, I., McDermid, J.: Generating and maintaining a safety argument for integrated modular systems. In: Proceedings of the 5th Australian Workshop on Industrial Experience with Safety Critical Systems and Software, pp. 31–41 (2000)
35. Prokhorova, Y., Laibinis, L., Troubitsyna, E.: Facilitating construction of safety cases from formal models in Event-B. Inf. Soft. Technol. **60**, 51–76 (2015)
36. Retouniotis, A., Papadopoulos, Y., Sorokos, I., Parker, D., Matragkas, N., Sharvia, S.: Model-connected safety cases. In: Bozzano, M., Papadopoulos, Y. (eds.) IMBSA 2017. LNCS, vol. 10437, pp. 50–63. Springer, Cham (2017). https://doi.org/10.1007/978-3-319-64119-5_4
37. Sandro, A.D., Selim, G.M.K., Salay, R., Viger, T., Chechik, M., Kokaly, S.: MMINT-A 2.0: tool support for the lifecycle of model-based safety artifacts. In: Proceedings of the 23rd International Conference on Model Driven Engineering Languages and Systems, pp. 15:1–15:5. ACM (2020)
38. Sljivo, I., Gallina, B., Carlson, J., Hansson, H.: Using safety contracts to guide the integration of reusable safety elements within ISO 26262. In: Proceedings of the 21st Pacific Rim International Symposium on Dependable Computing - PRDC, pp. 129–138. IEEE Computer Society (2015)
39. I. Standard. 26262: Road vehicles - functional safety. ISO (2018)
40. Törner, F., Öhman, P.: Automotive safety case a qualitative case study of drivers, usages, and issues. In: Proceedings of the 11th High Assurance Systems Engineering Symposium, pp. 313–322. IEEE Computer Society (2008)
41. UNDERWRITERS LABORATORIES INC.: ANSI/UL-4600 Standard for Evaluation of Autonomous Products (2020)
42. Vierhauser, M., et al.: Interlocking safety cases for unmanned autonomous systems in shared airspaces. Trans. Softw. Eng. **47**, 899–918 (2019)
43. Wardziński, A., Jones, P.: Uniform model interface for assurance case integration with system models. In: Tonetta, S., Schoitsch, E., Bitsch, F. (eds.) SAFECOMP 2017. LNCS, vol. 10489, pp. 39–51. Springer, Cham (2017). https://doi.org/10.1007/978-3-319-66284-8_4
44. Wohlin, C.: Guidelines for snowballing in systematic literature studies and a replication in software engineering. In: Proceedings of the 18th International Conference on Evaluation and Assessment in Software Engineering, pp. 1–10. ACM (2014)
45. Zhang, H., Babar, M.A., Tell, P.: Identifying relevant studies in software engineering. Inf. Softw. Technol. **53**(6), 625–637 (2011)

Towards Certified Analysis of Software Product Line Safety Cases

Ramy Shahin[1]([✉]), Sahar Kokaly[2], and Marsha Chechik[1]

[1] University of Toronto, Toronto, Canada
{rshahin,chechik}@cs.toronto.edu
[2] General Motors, Markham, Canada
sahar.kokaly@gm.com

Abstract. Safety-critical software systems are in many cases designed and implemented as families of products, usually referred to as Software Product Lines (SPLs). Products within an SPL vary from each other in terms of which features they include. Applying existing analysis techniques to SPLs and their safety cases is usually challenging because of the potentially exponential number of products with respect to the number of supported features. In this paper, we present a methodology and infrastructure for certified *lifting* of existing single-product safety analyses to product lines. To ensure certified safety of our infrastructure, we implement it in an interactive theorem prover, including formal definitions, lemmas, correctness criteria theorems, and proofs.

We apply this infrastructure to formalize and lift a Change Impact Assessment (CIA) algorithm. We present a formal definition of the lifted algorithm, outline its correctness proof (with the full machine-checked proof available online), and discuss its implementation within a model management framework.

Keywords: Safety cases · Product lines · Lean · Certified analysis

1 Introduction

The development of safety-critical systems usually involves a rigorous safety engineering process. A primary artifact resulting from that is a *safety case*, identifying potential safety hazards, their mitigation goals, and pieces of evidence required to show that goals have been achieved. Safety cases, together with other system artifacts, are usually inspected and analyzed by tools as a part of the safety engineering process. In safety-critical domains, correctness of those tools is essential to the integrity of the whole process. *Correctness certification* of tools w.r.t. their specifications becomes of extremely high value in this context.

In many cases, families of safety-critical software products are developed together in the form of *Software Product Lines (SPLs)*. Different product variants of an SPL have different *features*, i.e., externally visible attributes such as a piece of functionality, support for a particular peripheral device, or a performance

© Springer Nature Switzerland AG 2021
I. Habli et al. (Eds.): SAFECOMP 2021, LNCS 12852, pp. 130–145, 2021.
https://doi.org/10.1007/978-3-030-83903-1_9

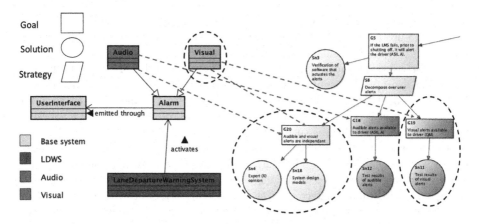

Fig. 1. Lifted change impact assessment when the "Visual" class is modified. A dashed ellipse around the Visual class denotes a modification, and dashed ellipses around safety case elements indicate that they need to be rechecked as a result.

optimization. Each feature can be either present or absent in each of the product variants of an SPL. Given this combinatorial nature of feature composition, analyzing the safety of each product instance individually in a *brute-force* fashion is usually intractable.

Several source-code and model-based analysis tools have been *lifted* to product lines [2,5,9,12,20,22–24] in the sense that they can be applied efficiently to the whole product line at once, leveraging the commonalities between individual products, and thus generating aggregated results for the complete product set. Those results have to be correct with respect to applying the analysis to each product individually. However, to the best of our knowledge, lifting of safety analyses has not been attempted before.

In this paper, we present a systematic methodology to correct-by-construction lifting of safety case analysis algorithms to software product lines. This includes infrastructure building blocks for implementing lifted algorithms, and proving their correctness with respect to their single-product counterparts. We use the Lean interactive theorem prover [18] to formalize the correctness criteria of lifting, implement our lifting infrastructure, and prove the correctness of lifted algorithms. A Lean proof is machine-checked, so it can be used as a *correctness certificate* of the property being proven.

We demonstrate our approach on a *Change Impact Assessment (CIA)* algorithm [15] that takes a system model, an assurance case, traceability links in between, and a modification to the system model as inputs, and determines the set of safety case elements that need to be revised or rechecked.

Motivating Example. Consider the *Lane Management System (LMS)* system outlined in [4]. LMS can be thought of as a product line with several features, including: *Lane Departure Warning System (LDWS)*, *Audio warning (Audio)*, and *Visual warning (Visual)*. For simplicity of presentation, we assume that all feature combinations are allowed. Figure 1 shows a snippet of the class diagram of the LMS product line, and the corresponding snippet of its GSN [13] assurance case, with traceability links in between the two.

We use colored annotations to map class diagram and GSN elements to features. For example, elements colored in green belong to the `Audio` feature, and those in orange belong to the `LDWS` feature. Base system elements (existing in all products) are in yellow. In general each element can be annotated by a propositional formula over features (usually referred to as a *presence condition*).

Consider a modification to the `Visual` class. The problem CIA algorithms try to solve is figuring out how that modification of a system element would impact the safety case. We distinguish between two ways in which a change to the system can impact safety case elements [15]: (1) `revise` – the *content* of the element (e.g., definition of a goal, or description of a solution) may have to be revised because it referred to a system element that has changed and the semantics of the content may have changed, and (2) `recheck` – the *state* of the element (e.g., whether a goal is satisfied, or a solution is available) must be rechecked because it may have changed.

In a product line setting, in addition to figuring out which elements are impacted, we also need to identify the product variants in which they are. In Fig. 1, goals `G19` and `G20` are directly impacted by modifications to class `Visual` because of the direct traceability links. Both classes need to be rechecked as a result, but only in products where the `Visual` feature is included. In the same set of products, pieces of evidence linked to those goals (`Sn4`, `Sn11`, `Sn18`) need to be rechecked as well. Note that although `G20`, `Sn4`, and `Sn18` belong to all product variants, we do not need to recheck them in product variants not including the `Visual` feature.

A CIA tool lifted to product lines has to preserve the exact semantics of its single-product counterpart. In other words, using the lifted tool should output exactly the union of outputs of the single-product tool applied to each product variant. A software bug in the lifted tool might result in false positives (elements marked as impacted while they should not). Even worse, a bug might result in overlooking an impacted element, potentially resulting in safety incidents.

Contributions. In this paper, we (1) outline a methodology for lifting safety analyses to safety cases of software product lines, and present a generic infrastructure for certified lifting (data structures and correctness criteria) using the Lean interactive theorem prover; and (2) demonstrate our methodology on a CIA algorithm lifted to software product lines, i.e., supporting the input of feature-specific modifications, and outputting feature-specific annotations of safety case elements. In addition, (3) we formalize the single-configuration CIA algorithm from [15] using Lean; (4) we outline a sketch of the correctness proof of the lifted CIA algorithm with respect to the single-configuration one (full Lean proof available online); and (5) we discuss extending the MMINT-A [7] model management framework with lifted safety algorithms, including lifted CIA.

Organization. The rest of this paper is organized as follows: In Sect. 2, we provide background on safety cases and SPLs. We outline the correctness criteria, methodology, and infrastructure needed to formally lift safety case algorithms in Sect. 3. In Sect. 4 we formalize the original single-configuration CIA algorithm, its lifted counterpart, and outline the lifting correctness proof. Section 5 explains how lifted algorithms can be integrated into existing model management tools. Section 6 compares our approach to related work, and Sect. 7 concludes.

2 Background

In this section, we provide background on safety cases, the GSN notation, Change Impact Assessment (CIA), and Software Product Lines (SPLs).

2.1 Safety Cases, GSN, and Change Impact Assessment

A safety case is a structured argument, decomposing safety goals into sub-goals, and linking pieces of safety evidence to the goals. Safety goals are usually identified using hazard assessment techniques. Each of the hazards needs to be mitigated by fulfilling one or more safety goal(s).

Goal Structured Notation (GSN) [13] is a graphical notation for defining safety cases. The safety case portion of Fig. 1 is an example of a GSN safety case model. A GSN model has elements of four different types. A *goal* is either satisfied or not based on the states of its sub-goals, connected solution nodes, and the semantics of decomposition strategy nodes involved. A *solution* is a piece of evidence that needs to be validated for its connected goal(s) to be satisfied. A *strategy* is a decomposition of a goal into sub-goals. A *context* connected to a goal node adds contextual assumptions that are assumed to hold when evaluating whether a goal is satisfied or not.

GSN-IA [15] is an algorithm for reflecting changes made in system models onto the relevant GSN safety cases. The inputs to GSN-IA are the initial system model S and a safety case A connected by a traceability mapping R, the changed system S' and the delta D recording the changes between S and S'. Specifically, D is the triple $\langle C0a, C0d, C0m \rangle$ where $C0a$, $C0d$, and $C0m$ are the sets of elements added, deleted, and modified respectively. The output of GSN-IA is the annotated model K of the safety case A, indicating which elements are marked for revise, recheck, or reuse.

GSN-IA is parameterized by three slicers [21]: a system model slicer Slice_{Sys}, and two safety case slicers Slice_{GSN_V} and Slice_{GSN_R}. Slice_{Sys} is used to determine how the impact of modifications propagates within the system model. Similarly, the safety case slicers trace through dependencies within the safety case, with Slice_{GSN_V} only tracing direct dependencies, while Slice_{GSN_R} recursively generates the transitive closure of dependencies.

2.2 Software Product Lines

We introduce Software Product Line (SPL) concepts following definitions from [20]. An SPL $\mathcal{L}$ is a tuple (F, Φ, D, ϕ) where: (1) F is the set of features s.t.

an individual product can be derived from $\mathcal{L}$ via a *feature configuration* $\rho \subseteq F$. (2) $\Phi \in \text{Prop}(F)$ is a propositional formula over F defining the valid set of feature configurations. Φ is called a *Feature Model (FM)* of $\mathcal{L}$. The set of valid configurations defined by Φ is called $\text{Conf}(\mathcal{L})$. (3) D is a set of program elements, called the *domain model*. The whole set of program elements is usually referred to as the *150% representation* of $\mathcal{L}$. (4) $\phi : D \rightarrow \text{Prop}(F)$ is a total function mapping each program element to a proposition (*feature expression*) defined over the set of features F. $\phi(e)$ is called the *Presence Condition (PC)* of element e, i.e. the set of product configurations in which e is present.

Given a product line $\mathcal{L}$ and a feature configuration ρ, we define $\mathcal{L}|_\rho$ to be the subset of elements of $\mathcal{L}$ that belong to at least one of the features in ρ. We loosely use the same indexing operator when referring to subsets of values in a data structure subject to a feature configuration. For example, given a feature configuration {LDWS, Visual}, a product with all the elements, except for the ones annotated in green, is instantiated from the product line in Fig. 1.

3   Methodology and Infrastructure

In this section, we present a set of generic infrastructure building blocks that can be used in designing and certifying the correctness of variability-aware algorithms applied to safety cases. We then present the correctness criteria of variability-aware algorithms with respect to their single-product counterparts. Finally, we put the infrastructure together with the correctness criteria into a correct-by-construction methodology for systematic lifting of safety case algorithms. Infrastructure definitions, theorems, and proofs are implemented in 53 Lean lines of code.

We formalize algorithms, theorems, and proofs using the Lean [18] interactive theorem prover. We had two requirements for the proof assistant to be used in this project: (1) to be based on constructive rather than classical logic, to allow for explicit tracing of which sub-goals (and their proof evidence) contribute to the overall proof; and (2) to allow for sound user-defined proof automation procedures, which can reduce the human effort involved in the proof development process. Lean meets those two requirements. It is based on the Calculus of Inductive Constructions [1], so it supports constructive logic by default. It also supports tactic-based meta-programming of theorems and proof objects.

Lifted Data Structures. The types of all input, output, and intermediate data structures of an algorithm need to be lifted, i.e., elements of each of those data structures need to be paired with presence conditions, indicating the set of products this element belongs to. Listing 1 has definitions of some of the data types used for lifted data structures. PC (line 1) is the type for presence conditions, which is defined as native Lean propositions. Var (line 2) is a higher-order lifted type, taking a type α as a parameter, and pairing values of type α with presence conditions.

The lifted set data type set' (line 3) is a higher-order type parameterized by type α, and implemented as a function $\alpha \rightarrow \text{PC}$. This implementation happens to

```
1   def PC := Prop
2   structure Var (α) : Type) := (v : α) (pc : PC)
3   def set' (α : Type) : Type := α → PC
4   def index (s : set' α) (pc : PC) : set α := (and pc) ∘ s
5   def mem (x : Var α) (s : set' α) : Prop := x.pc → (s x.v)
6   def subset (s₁ s₂ : set' α) : Prop := ∀ a, mem a s₁ → mem a s₂
7   def union (s₁ s₂ : set' α) : set' α := λ x, (s₁ x) ∨ (s₂ x)
8   def image (f : α → β) (s : set' α) : set' β := λ x, (∀ y, f y = x ∧ s y)
```

Listing 1. Variability-aware building blocks.

$$\begin{array}{ccc} \mathcal{L} & \xrightarrow{f'} & R' \\ \downarrow{|\rho} & & \downarrow{|\rho} \\ P & \xrightarrow{f} & R \end{array} \qquad\qquad \begin{array}{ccccc} \mathcal{L} & \xrightarrow{f'} & R' & \xrightarrow{g'} & S' \\ \downarrow{|\rho} & & \downarrow{|\rho} & & \downarrow{|\rho} \\ P & \xrightarrow{f} & R & \xrightarrow{g} & S \end{array}$$

(a) Correctness of a lifted function [22]. (b) Correctness of lifted function composition.

Fig. 2. Lifting correctness criteria.

be the same as the implementation of Lean sets. However, the semantics of Lean sets assume that a value of type α is either present or absent in a set over α. Lifted sets on the other hand map an element of type α to an arbitrary propositional formula which might evaluate to True (i.e., the element exists in the set in all configurations), False (i.e., the element does not exist in any configuration of the set), or a contingent formula indicating the set of configurations in which the element exists in the set.

The primary operation on lifted data types in general is indexing. Given a lifted set s and a presence condition pc, index s pc evaluates to a Lean set (not lifted) of elements existing in the configurations satisfied by pc in s. This is exactly how the index operator is defined on set', conjoining pc with the presence condition of each element in s (line 4).

Standard set operations also need to be overloaded for lifted sets. Lifted set membership semantically checks if a lifted value (v,pc) exists in all configurations of a lifted set s. It is defined (line 5) as a propositional implication between pc (the set of configurations where the lifted value exists), and the set of configurations where v exists in s.

Lifted subset is defined exactly the same as standard subset, using the lifted definition of set membership (line 6). Similarly, lifted set union is implemented as a disjunction of the propositional definitions of its two arguments (line 7). The last lifted set operation is image (line 8), taking a function $f : \alpha \to \beta$ and a lifted set s of α, and applies f to each element s, returning a lifted set of β.

Correctness Criteria. Given a product line $\mathcal{L}$, an analysis algorithm f, and a product configuration ρ, we construct a lifted version of f (referred to as f'), such that instantiating a product P from $\mathcal{L}$ using configuration ρ, and then applying f to P has the same result as applying f' to $\mathcal{L}$ and then instantiating a

```
1   variables (f  : set  α → set  β) (g  : set  β  → set  γ)
2   variables (f′ : set′ α → set′ β) (g′ : set′ β  → set′ γ)
3   theorem fun_comp_correct :
4      (∀ a ρ, (f (a | ρ) = (f′ a) | ρ)) → (∀ b ρ, (g (b | ρ) = (g′ b) | ρ)) →
5      (∀ a ρ, (g ∘ f) (a | ρ) = ((g′ ∘ f′) a) | ρ)
```

Listing 2. Lifted function composition theorem.

product-specific result using ρ. This is summarized by the commuting diagram in Fig. 2a [22].

Lifting Methodology. We follow a divide-and-conquer methodology to design lifted analyses from their single-product counterparts. If an analysis algorithm is broken-down into smaller functions, and each of those functions is individually lifted, composing the lifted functions together has to preserve the correctness criteria. This is summarized in Fig. 2b.

We formulate the correctness criteria of lifted function composition as a theorem (Listing 2). Assume we have two functions (f:set α → set β) and (g:set β → set γ), and their two lifted functions (f′:set′ α → set′ β) and (g′:set′ β → set′ γ) respectively (Line 2). The theorem states that if f′ is a correct lifting of f, and g′ is a correct lifting of g, then g′ ∘ f′ is a correct lifting of g ∘ f (Lines 3-5). The theorem is proven by term rewriting. Definitions of all theorems, lemmas, and their full Lean proofs are available online[1].

Correctness of the lifted function composition theorem is the foundation of compositional lifting correctness proofs. Small helper functions can be manually lifted and proven correct relatively easily, and their correctness proofs can be composed together with composing the functions themselves using the theorem. This way, lifted analyses can be compositionally implemented following the same structure of their single-product counterparts, composing correctness proofs together with function composition. We demonstrate this methodology on lifting a Change Impact Assessment (CIA) algorithm in Sect. 4.

4   Change Impact Assessment

In this section, we formalize the GSN-IA [15] impact assessment algorithm, systematically design a lifted version of it, and prove its correctness based on the methodology in Sect. 3. In total, the GSN-IA algorithm is formalized in 45 Lean lines of code. The formalization of the lifted algorithm, together with correctness lemmas and proofs takes 160 lines of Lean code.

4.1   Single-Product Algorithm

The data types and external dependencies of the GSN_IA algorithm are defined in Listing 3. Annotation is the data type of annotations assigned to GSN model elements, with the values Reuse, Recheck, and Revise (lines 1-2). SysEl and GSNEl are opaque types of system model elements and GSN model elements

---

[1] https://github.com/ramyshahin/variability.

```
1   inductive Annotation : Type
2   | Reuse | Recheck | Revise
3
4   constants SysEl GSNEl : Type
5   def Sys      : Type := set SysEl
6   def GSN      : Type := set GSNEl
7   def TraceRel : Type := set (SysEl × GSNEl)
8
9   variable sliceSys   (s : Sys)  (es : set SysEl) : Sys
10  variable sliceGSN_V (ac : GSN) (es : set GSNEl) : GSN
11  variable sliceGSN_R (ac : GSN) (es : set GSNEl) : GSN
12  structure Delta := (add: set SysEl) (delete: set SysEl) (modify: set SysEl)
```

Listing 3. Type definitions of the formalized GSN_IA algorithm.

respectively, where a system model `Sys` and a GSN model `GSN` are sets of each of those elements types (lines 4-6). `TraceRel` is a traceability relation between system model elements and GSN model elements, so it is a defined as a set of ordered pairs of `SysEl` and `GSNEl` (line 7). GSN_IA is parameterized by three model slicers: `sliceSys` is a system model slicer, while `sliceGSN_V` and `sliceGSN_R` are GSN model slicers. Each of the slicers takes a model and a set of elements used as the slicing criterion, returning a subset slice of the input model (lines 9–11). `Delta` is composed of three sets of system elements, representing the elements added, modified and deleted (lines 12). Listing 4 has the definitions of the GSN_IA algorithm, together with three helper functions. `restrict` is a function taking a traceability relation `t` and a delta `es` as inputs, and returns a restricted subset of `t` only covering elements in `es` (lines 1–2). `trace` takes a traceability relation `t` and a set of system elements `es` as inputs, and returns the set of GSN elements mapped from `es` by `t` (lines 4–5). `createAnnotation` assigns an `Annotation` value to each element in a GSN model, given sets of elements to be rechecked and revised (lines 7–12).

The change impact assessment algorithm GSN_IA takes two system models `S` and `S'` and the delta `D` between them. It also takes a GSN model `A` and a traceability relation `R` between system model elements and GSN model elements. It returns a set of ordered pairs of GSN model elements and annotations. The algorithm starts by restricting the traceability relation based on `D`, slices the original system model `S` using the elements deleted and modified as a slicing criterion, and slices the modified system model `S'` using the added and modified elements as the slicing criterion (lines 16–18). Using those two slices, the corresponding GSN model elements are traced using the traceability relation (line 19). The GSN elements traced from elements deleted from the original system model are to be revised (line 20). The slice of the GSN model based on the traced elements are to be rechecked (lines 21–22), and both revise and recheck sets are used to annotate the GSN model elements (line 23).

```
1   def restrict (t : TraceRel) (d : Delta) : TraceRel :=
2       λ x, x.1 ∈ d.add ∪ d.delete ∪ d.modify
3
4   def trace (t : TraceRel) (es : set SysEl) : set GSNEl :=
5       image prod.snd {p | p ∈ t ∧ p.1 ∈ es}
6
7   def createAnnotation (g : GSN) (recheck : set GSNEl) (revise : set GSNEl)
8           : set (GSNEl × Annotation) :=
9       let ch := image (λ e, (e, Annotation.Recheck)) recheck,
10          rv := image (λ e, (e, Annotation.Revise)) revise,
11          ru := image (λ e, (e, Annotation.Reuse)) (g − (recheck ∪ revise))
12      in ch ∪ rv ∪ ru
13
14  def GSN_IA (S : Sys) (S' : Sys) (A : GSN) (R : TraceRel) (D : Delta)
15          : set (GSNEl × Annotation) :=
16      let R'          := restrict R D,
17          C1dm        := sliceSys S ((delete D) ∪ (modify D)),
18          C1am        := sliceSys S' ((add D) ∪ (modify D)),
19          C2Recheck   := (trace R C1dm) ∪ (trace R' C1am),
20          C2Revise    := trace R (delete D),
21          C3Recheck1  := sliceGSN_V A C2Revise,
22          C3Recheck2  := sliceGSN_R A (C2Recheck ∪ C3Recheck1)
23      in  createAnnotation A C3Recheck2 C2Revise
```

Listing 4. Helper functions and the formalized GSN_IA algorithm.

4.2 Lifted Algorithm

Listing 5 is the variability-aware version of the algorithm in Listing 4. Both algorithms are compositions of function/operator calls, so each of those functions/operators is replaced with its lifted counterpart. We assume that lifted versions of the three slicers are provided, and that they meet the correctness criteria of Fig. 2a.

All the set types used in GSN_IA need to be lifted. Definitions in lines 1–4 are lifted sets of system model elements, GSN model elements, and traceability mappings. A lifted delta (line 4) is composed of three lifted sets (additions, deletions and modifications).

The proof of the correctness theorem used auxiliary correctness lemmas for each of the helper algorithms. Each of the proofs expands definitions and repeatedly applies the correctness of lifted function composition (Fig. 2b).

Lifted Helper Algorithms. Since the lifted CIA algorithm operates on lifted data structures, all helper algorithms need to be modified to correctly operate on lifted data structures as well. In particular, we outline lifted versions of restrict and trace (Listing 6).

The original implementation of restrict takes a traceability map and a delta as inputs, and returns the minimal subset of the traceability map that covers all the elements in the delta. We now have presence conditions associated to system model elements, assurance case elements, and also the traceability links

```
1   def Sys' : Type := set' SysEl
2   def GSN' : Type := set' GSNEl
3   def TraceRel' : Type := set' (SysEl × GSNEl)
4   structure Delta' :=(add: set' SysEl)(delete: set' SysEl)(modify: set' SysEl)
5
6   def GSN_IA' (S S' : Sys') (A  : GSN') (R  : TraceRel') (D  : Delta')
7        : set' (GSNEl × Annotation) :=
8    let R'         := restrict' R D,
9        C1dm        := sliceSys' S (D.delete ∪ D.modify),
10       C1am        := sliceSys' S' (D.add ∪ D.modify),
11       C2Recheck   := (trace' R C1dm) ∪ (trace' R' C1am),
12       C2Revise    := trace' R D.delete,
13       C3Recheck1  := sliceGSN_V' A C2Revise,
14       C3Recheck2  := sliceGSN_R' A (C2Recheck ∪ C3Recheck1)
15    in createAnnotation' A C3Recheck2 C2Revise
```
Listing 5. Lifted Change Impact Assessment algorithm.

```
1   def restrict' (t : TraceRel') (d : Delta') : TraceRel' :=
2     let relevant := d.add ∪ d.delete ∪ d.modify
3     in λ x, t x ∧ relevant x.1
4
5   def trace' (t : TraceRel') (es : set' SysEl) : set' GSNEl :=
6     λ (g:GSNEl), ∃ (s : SysEl) , es s ∧ t ⟨s , g⟩
```
Listing 6. Lifted implementation of restrict and trace.

in between. The lifted version of `restrict` (referred to as `restrict'`) needs to correctly process all those presence conditions.

The lifted algorithm starts by calculating the set of relevant elements in the system model, which is the union of added, deleted and modified elements in the delta (line 2). The algorithm returns a lifted traceability mapping as a function taking `((s,g),pc)`, where `(s,g)` is a system model element-GSN model element pair, and `pc` is a presence condition. This function evaluates to the conjunction of applying the input traceability map `t` to `((s,g),pc)`, and applying `relevant` to `(s,g)`. Recall that variability-aware sets (as well as Lean sets) are functions mapping values of a given type to propositions.

Similarly, `trace'` is the lifted version of `trace`. The returned lifted set is a function mapping a GSN model element `g` to the set of configurations from which there exists a system model element `s` in the input lifted set of system elements, where `(s,g)` belongs to the input traceability map.

The lifted version of `createAnnotation` (named `createAnnotation'`) is of exactly the same structure as the original because it strictly uses set operations (union, set difference and image), which have been all lifted as a part of the underlying variability-aware set implementation (Listing 1). The correctness theorem of GSN_IA' with respect to GSN_IA is in Listing 7. It is a direct instantiation of the general correctness criteria in Fig. 2a, applied to inputs of the GSN_IA algorithm.

```
1   theorem GSN_IA'_correct :
2     ∀ (S S' : Sys') (A  : GSN') (R  : TraceRel') (D : Delta') (pc : PC),
3     (GSN_IA' S S' A R D) | pc = GSN_IA  (S | pc) (S' | pc) (A | pc) (R | pc) (D | pc)
```
Listing 7. Correctness theorem of GSN_IA'.

4.3 Examples

$A = \{(G5, \text{TT}), (Sn3, \text{TT}), (S8, \text{TT}), (G18, \text{Audio}), (G19, \text{Visual}), (G20, \text{True}),$
$\quad (Sn4, \text{TT}), (Sn18, \text{TT}), (Sn12, \text{Audio}), (Sn11, \text{Visual})\}$
$R = \{(Visual, G19, \text{Visual}), (Visual, G20, \text{Visual}), (Audio, G18, \text{Audio}),$
$\quad (Audio, G20, \text{Audio})\}$

(a) Assurance case elements (A) and traceability links (R) used in Ex1 and Ex2.

$S = \{(Alarm, \text{True}), (UserInterface, \text{True}), (Audio, \text{Audio}), (Visual, \text{Visual}),$
$\quad (LaneDepartureWarningSystem, \text{LDWS})\}$
$S' = \{(Alarm, \text{True}), (UserInterface, \text{True}), (Audio, \text{Audio}), (Visual', \text{Visual}),$
$\quad (LaneDepartureWarningSystem, \text{LDWS})\}$
$D = \langle\{\}, \{\}, \{(Visual, \text{Visual})\}\rangle$

(b) System model (S), modified system model (S'), and delta (D) used in Ex1.

$S = \{(Alarm, \text{True}), (UserInterface, \text{True}), (Audio, \text{Audio}), (Visual, \text{Visual}),$
$\quad (LaneDepartureWarningSystem, \text{LDWS})\}$
$S' = \{(Alarm', \text{True}), (UserInterface, \text{True}), (Audio, \text{Audio}), (Visual, \text{Visual}),$
$\quad (LaneDepartureWarningSystem, \text{LDWS})\}$
$D = \langle\{\}, \{\}, \{(Alarm, \text{True})\}\rangle$

(c) System model (S), modified system model (S'), and delta (D) used in Ex2.

Fig. 3. Inputs to the GSN-IA' algorithm used in Ex1 and Ex2.

In this section, we apply our lifted CIA algorithm to two examples of modifications to the fragment of the LMS product line presented in Sect. 1 (Fig. 1).

Ex1: Feature-Specific Modification. Suppose that the *Visual* class is modified. This class is local to the Visual feature. If we only analyze the fragment in Fig. 1, the inputs to GSN-IA' are shown in Fig. 3a and Fig. 3b.

Tracing through the algorithm, the first step is using restrict' to calculate R' = {(Visual, G19, Visual), (Visual, G20, Visual)} (line 8). Because C0a and C0d are both empty, and assuming a backward slicer (returning the transitive closure of the elements that might affect the slicing criteria),

C1dm and C1am both become {(Alarm, True), (Visual, Visual), (LaneDeparture WarningSystem, LDWS)} (lines 9–10). Now tracing from C1dm and C1am, C2recheck becomes {(G19, Visual), (G20, Visual)} (line 11). Since C0d is empty, C2revise and C3recheck1 are both empty as well (lines 12–13). Using a backward GSN slicer, C3recheck2 becomes {(G19, Visual), (G20, Visual), (Sn11, Visual), (Sn4, Visual), (Sn18, Visual)} (line 14). The algorithm returns an empty set of GSN elements to be revised, and the set C3recheck2 to be rechecked. Note that G20, Sn4, and Sn18 are all base model elements (having True as a presence condition), so the algorithm output states that we need to recheck those elements only in products where the feature Visual is present.

Ex2: Base System Modification. Suppose that the *Alarm* class is modified. This is a base system class, i.e., it is present in all products. The inputs to GSN-IA' (restricted to the fragment in Fig. 1) are shown in Fig. 3a and Fig. 3c.

Since the Alarm class does not have any direct traceability links, R' is empty (line 8). Using a backward slicer (like in Ex1), C1dm and C1am both become {(Alarm, True), (Visual, Visual), (Audio, Audio), (LaneDepartureWarningSystem, LDWS)} (lines 9–10). From C1dm and C1am using the traceability links, C2recheck becomes {(G18, Audio), (G19, Visual), (G20, Visual)} (line 11). Again, since C0d is empty, C2revise and C3recheck1 are both empty as well (lines 12–13). With a backward GSN slicer, C3recheck2 becomes {(G18, Audio), (G19, Visual), (G20, Visual ∨ Audio), (Sn11, Visual), (Sn12, Audio), (Sn4, Visual ∨ Audio), (Sn18, Visual ∨ Audio)} (line 14). The algorithm returns an empty set of GSN elements to be revised, and the set C3recheck2 to be rechecked. Note that in this example, G20, Sn4, and Sn18 are annotated with recheck with presence condition Visual ∨ Audio, which means that they need to be rechecked only if either Audio or Visual are present.

5   Towards Implementation

The GSN-IA algorithm is implemented, together with slicers and model operators, as an extension of the MMINT [6] model management framework (Fig. 4), called MMINT-A [7]. In order to extend MMINT-A to support annotative product line models, and subsequently the lifted change impact assessment algorithm, the following modifications are required: (1) Model elements need to be extended with presence conditions, with True as a default value. This way single product models (where all elements have the default True presence condition) are directly supported as well. (2) Operators on models need to be modified to take presence conditions into consideration, and compute the presence conditions of their outputs. Those modifications are mostly systematic along the lines of those of restrict' and trace' (Listing 6). (3) Higher-level algorithms (e.g., GSN-IA) need to be modified accordingly to use the lifted versions of the operators. (4) The user interface of MMINT-A needs to support annotating different model elements with presence conditions. (5) Optionally, MMINT-A can check the well-formedness of presence condition annotations. For example, the presence condition of an association between two UML classes has to be subsumed by the presence conditions of its two end points.

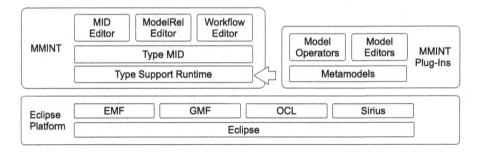

Fig. 4. Architecture of the MMINT model management framework [7].

6   Related Work

Model-Based Approaches to Safety Case Management. Many methods for modeling safety cases have been proposed, including goal models and requirements models [3,10] and GSN [13]. The latter is arguably the most widely used model-based approach to improving the structure of safety arguments. Building on GSN, Habli et al. [11] examine how model-driven development can provide a basis for the systematic generation of functional safety requirements and demonstrates how an automotive safety case can be developed. Gallina [8] proposes a model-driven safety certification method to derive arguments as goal structures given in GSN from process models. The process is illustrated by generating arguments in the context of ISO 26262. We consider this category of work complimentary to ours; we do not focus on safety case construction but instead assume presence of a safety case and focus on assessing the impact of system changes.

Lifting to Software Product Lines. Different kinds of software analyses have been re-implemented to support product lines [24]. For example, the TypeChef project implements variability aware parsers and type checkers for Java and C [12]. The SuperC project [9] is another C language variability-aware parser. A graph transformation engine was lifted to product lines of graphs [20]. Datalog-based analyses (e.g., pointer analysis) have been lifted by modifying the Datalog engine being used [23]. SPL^{Lift} [2] lifts data flow analyses to annotative product lines. Model checkers based on Featured Transition Systems [5] check temporal properties of transition systems where transitions can be labeled by presence conditions. Syntactic transformation techniques have been suggested for lifting abstract interpretation analyses [17] and functional analyses [22] to SPLs.

In this paper, our methodology tailors the lifting approach from related work to safety cases of product lines, and we demonstrate it on change impact assessment. We tackle a new class of product line artifacts, particularly safety cases.

To the best of our knowledge, this is the first attempt to lift a safety case analysis to product lines.

Formalized Systems and Interactive Theorem Proving. Correctness and behavioral properties of several software systems have been formalized and verified using interactive theorem provers. The CompCert compiler [16] is an example of a C-language compiler fully certified using the Coq theorem prover. The seL4 microkernel [14] was verified using the Isabelle\HOL theorem prover. Isabelle was also used to formalize the Structured Assurance Case Metamodel (SACM) notation for certified definition of assurance cases [19].

7   Conclusion and Future Work

In this paper, we presented a methodology for lifting safety case analysis algorithms to software product lines. We also outlined a certification infrastructure (data structures and correctness criteria) for our lifting approach using the Lean interactive theorem prover. We demonstrated both the approach and correctness certification on formalizing and lifting a Change Impact Assessment (CIA) algorithm [15]. We discussed the implementation of the lifted CIA algorithm as part of the safety model management system MMINT-A [7]. A lifted CIA algorithm allows for reusing impact assessment conclusions across a potentially exponential (in number of features) different product variants, as opposed to using a product-level CIA algorithm in individual product instances, which is intractable in most cases.

For future work, we are working together with an industrial partner on applying our lifted algorithm to their assurance case models. We also plan to lift other safety case algorithms (including slicers), and add their implementations to MMINT-A. Visualization of the analysis results and improved user interaction is another area of future improvements.

Acknowledgments. We thank Rick Salay and members of the Software Modeling Group at University of Toronto for discussions and suggestions throughout this project. We also thank the anonymous reviewers for their insightful feedback. This work was supported by General Motors and NSERC.

References

1. Bertot, Y., Castran, P.: Interactive Theorem Proving and Program Development: Coq'Art The Calculus of Inductive Constructions, 1st edn. Springer Publishing Company, Incorporated, Berlin (2010)
2. Bodden, E., Tolêdo, T., Ribeiro, M., Brabrand, C., Borba, P., Mezini, M.: SPLLIFT: statically analyzing software product lines in minutes instead of years. In: Proceedings of 2013 ACM Conference on Programming Language Design and Implementation (PLDI 2013), pp. 355–364. ACM (2013)
3. Brunel, J., Cazin, J.: Formal verification of a safety argumentation and application to a complex UAV system. In: Proceedings of 31st International Conference on Computer Safety, Reliability, and Security (SAFECOMP 2012) (2012)

4. Chechik, M., Kokaly, S., Rahimi, M., Salay, R., Viger, T.: Uncertainty, modeling and safety assurance: towards a unified framework. In: Chakraborty, S., Navas, J.A. (eds.) VSTTE 2019. LNCS, vol. 12031, pp. 19–29. Springer, Cham (2020). https://doi.org/10.1007/978-3-030-41600-3_2

5. Classen, A., Cordy, M., Schobbens, P.Y., Heymans, P., Legay, A., Raskin, J.F.: Featured transition systems: foundations for verifying variability-intensive systems and their application to LTL Model checking. IEEE Trans. Softw. Eng. **39**(8), 1069–1089 (2013)

6. Di Sandro, A., Salay, R., Famelis, M., Kokaly, S., Chechik, M.: MMINT: a graphical tool for interactive model management. In: Proceedings of MODELS (Demo) (2015)

7. Fung, N.L.S., Kokaly, S., Di Sandro, A., Salay, R., Chechik, M.: MMINT-A: A Tool for Automated Change Impact Assessment on Assurance Cases. In: Proceedings 37th International Conference on Computer Safety, Reliability, and Security (SafeComp 2018), pp. 60–70 (2018)

8. Gallina, B.: A model-driven safety certification method for process compliance. In: Proceedings of EEE 25th International Symposium on Software Reliability Engineering (ISSRE 2014), pp. 204–209. IEEE (2014)

9. Gazzillo, P., Grimm, R.: SuperC: parsing all of C by taming the preprocessor. In: Proceedings of the 33rd ACM SIGPLAN Conference on Programming Language Design and Implementation (PLDI 2012), pp. 323–334. ACM (2012)

10. Ghanavati, S., Amyot, D., Peyton, L.: A Systematic review of goal-oriented requirements management frameworks for business process compliance. In: Proceedings of EEE International Workshop on Requirements Engineering and Law (RELAW 2011), pp. 25–34. IEEE (2011)

11. Habli, I., Ibarra, I., Rivett, R.S., Kelly, T.: Model-Based Assurance for Justifying Automotive Functional Safety. Tech. Report, SAE (2010)

12. Kästner, C., Apel, S., Thüm, T., Saake, G.: Type checking annotation-based product lines. ACM Trans. Softw. Eng. Methodol. **21**(3), 14:1–14:39 (2012)

13. Kelly, T., Weaver, R.: The goal structuring notation - a safety argument notation. In: International Conference on Dependable Systems and Networks (DSN 2004) (2004)

14. Klein, G., et al.: SeL4: formal verification of an OS kernel. In: Proceedings of the ACM SIGOPS 22nd Symposium on Operating Systems Principles, pp. 207–220 (2009)

15. Kokaly, S., Salay, R., Chechik, M., Lawford, M., Maibaum, T.: Safety case impact assessment in automotive software systems: an improved model-based approach. In: Tonetta, S., Schoitsch, E., Bitsch, F. (eds.) SAFECOMP 2017. LNCS, vol. 10488, pp. 69–85. Springer, Cham (2017). https://doi.org/10.1007/978-3-319-66266-4_5

16. Leroy, X.: A formally verified compiler back-end. J. Autom. Reason. **43**(4), 363–446 (2009). http://xavierleroy.org/publi/compcert-backend.pdf

17. Midtgaard, J., Dimovski, A.S., Brabrand, C., Wąsowski, A.: Systematic derivation of correct variability-aware program analyses. Sci. Comput. Program. **105**(C), 145–170 (2015)

18. de Moura, L., Kong, S., Avigad, J., van Doorn, F., von Raumer, J.: The lean theorem prover (system description). In: Felty, A.P., Middeldorp, A. (eds.) CADE 2015. LNCS (LNAI), vol. 9195, pp. 378–388. Springer, Cham (2015). https://doi.org/10.1007/978-3-319-21401-6_26

19. Nemouchi, Y., Foster, S., Gleirscher, M., Kelly, T.: Isabelle/SACM: computer-assisted assurance cases with integrated formal methods. In: Ahrendt, W., Tapia Tarifa, S.L. (eds.) IFM 2019. LNCS, vol. 11918, pp. 379–398. Springer, Cham (2019). https://doi.org/10.1007/978-3-030-34968-4_21

20. Salay, R., Famelis, M., Rubin, J., Di Sandro, A., Chechik, M.: Lifting model transformations to product lines. In: Proceedings of 34th International Conference on Software Engineering ... Publication (ICSE2014). ACM, NY, USA (2014)

21. Salay, R., Kokaly, S., Chechik, M., Maibaum, T.: Heterogeneous Megamodel Slicing for Model Evolution. In: Proceedings of International Conference on Model Driven Engineering Languages and Systems, 2016 (ME@MoDELS 2016), pp. 50–59 (2016)

22. Shahin, R., Chechik, M.: Automatic and efficient variability-aware lifting of functional programs. Proc. ACM Program. Lang. 4(OOPSLA) (2020)

23. Shahin, R., Chechik, M., Salay, R.: Lifting datalog-based analyses to software product lines. In: Proc. of ACM Joint Meeting on European Software Engineering Conference and Symposium (ESEC/FSE 2019). ACM, New York, NY, USA (2019)

24. Thüm, T., Apel, S., Kästner, C., Schaefer, I., Saake, G.: A classification and survey of analysis strategies for software product lines. ACM Comput. Surv. 47(1), 6:1–6:45 (2014)

Machine Learning Applications

Safety Assurance of Machine Learning for Chassis Control Functions

Simon Burton[1], Iwo Kurzidem[1(✉)], Adrian Schwaiger[1], Philipp Schleiss[1], Michael Unterreiner[2], Torben Graeber[2], and Philipp Becker[2]

[1] Fraunhofer IKS, 80686 Munich, Germany
{simon.burton,iwo.kurzidem,adrian.schwaiger,
philipp.schleiss}@iks.fraunhofer.de
[2] Porsche AG, 71287 Weissach, Germany
{michael.unterreiner,torben.graeber,philipp.becker}@porsche.de

Abstract. This paper describes the application of machine learning techniques and an associated assurance case for a safety-relevant chassis control system. The method applied during the assurance process is described including the sources of evidence and deviations from previous ISO 26262 based approaches. The paper highlights how the choice of machine learning approach supports the assurance case, especially regarding the inherent explainability of the algorithm and its robustness to minor input changes. In addition, the challenges that arise if applying more complex machine learning technique, for example in the domain of automated driving, are also discussed. The main contribution of the paper is the demonstration of an assurance approach for machine learning for a comparatively simple function. This allowed the authors to develop a convincing assurance case, whilst identifying pragmatic considerations in the application of machine learning for safety-relevant functions.

Keywords: Assurance case · Safety engineering · Machine learning · Automotive software

1 Introduction

Recent advances in Machine Learning (ML) have demonstrated the potential for efficient and sophisticated classifications based on data-driven models [16]. This is especially visible in domains where conventional programming is difficult and computationally expensive. However with the increased application of ML techniques to safety-related tasks, concerns related to the probability of incorrect or inaccurate predictions have also increased. Current safety-related challenges in ML include, but are not limited to: explainability of decision-making, unreliable confidence information, inadequate approximations via limited data-sets, insufficient or incomplete definitions, and meaningful safety metrics [15]. These functional insufficiencies and safety concerns are especially important for ML in automated driving applications, as they may potentially impact the overall vehicle's safety goals [3]. As such, industry safety standards, such as, ISO

I. Habli et al. (Eds.): SAFECOMP 2021, LNCS 12852, pp. 149–162, 2021.
https://doi.org/10.1007/978-3-030-83903-1_10

26262 (Road vehicles – Functional safety) [6] and ISO/PAS 21448 (Safety of the Intended Functionality - SOTIF) [7] apply. However, while these ISO standards tackle operational safety and offer guidance for safety analyses, neither standard offers a complete and coherent safety assurance approach suited for ML [1]. These shortcomings are not just limited to automated driving functions alone, guidance in the standards is also lacking when applying ML to other classes of vehicle functions, such as powertrain and chassis control. Such functions can directly impact the stability of the vehicle and therefore contribute to vehicle-level safety goals. Hence, a comprehensive and tailored safety assurance is vital, before deployment of ML-based systems, to guarantee safety.

This paper demonstrates such an assurance approach for a road surface estimation based on sound patterns. Within the application, acoustic sensors are used to categorise road conditions between the classes of dry (*dry*) and not dry (*!dry*). This information is then used to adapt chassis control functions to the road surface traction. A misclassification of the surface condition could therefore lead to a hazardous control action.

The paper is organised as follows: first, an overview of related publications and ideas is given in Sect. 2, followed by a description of the case study in Sect. 3 and an outline of the proposed approach in Sect. 4. In Sect. 5 properties of the chosen ML technique are analysed with respect to their strengths and weaknesses for assuring safety. The insights of the analysis are discussed and summarised as lessons learned in Sect. 6, leading to a conclusion in Sect. 7.

2   Related Work

Safety standards already exist for automotive functionality. ISO 26262 focuses on functional safety and provides comprehensive guidelines for the analysis of conventional software and hardware failures and ISO/PAS 21448 addresses insufficiencies and potential exploits for (conventional) software and ML, such as performance limitations, impact from the environment and foreseeable misuse by third parties. However, both standards do not offer a general strategy or approach for validating safety of non-conventional SW, such as ML algorithms.

In [12] both ISO standards are combined to create a product development process for ML. The authors incorporate ISO/PAS 21448 into ISO 26262 work products and development phases. The proposed approach is heavily based on ISO 26262 definitions, such as the chapter enumeration, and workflow, e.g. V-model. The ML specific work products are handled as additional documentation within each development phase. However, open questions regarding applicability for complex systems, *semantic gaps* (cf. Sect. 4) and meaningful evidence acquisition still remain. Additionally, no examples or methods are given on how to generate these additional documents, as no case study is presented.

A different take on this matter is introduced by Picardi et al. in [11], with *argument patterns* to demonstrate the safety of ML. The presented safety assurance patterns are tailored for ML components and highlight *how* the collected evidence, assumptions, strategies and claims relate to overall system safety goals.

The patterns are graphically represented using GSN [14] and show how performance evidence is (indirectly) connected to specific ML safety requirements. The result is to coherently and unambiguously represent a compelling safety argumentation. Additionally, the authors show how the argumentation patterns can be applied within different stages and activities of a complete ML assurance process. In a further work, Picardi et al. utilises the argument patterns to develop an assurance case specifically for a ML component within medical diagnostics [10].

Outside the automotive domain, audio interpretation via ML, for instance in form of speech recognition, achieves impressive performance. However, most considerations, analyses and evaluations of ML actually exclude safety as a major desideratum [2].

In this paper, we apply similar argument patterns to [11] for an assurance case of a chassis control function based on ML. We point out similarities to and extensions of the ISO standards within our assurance case. Furthermore, we highlight how a suitable selection of a ML paradigm can support the assurance case claims.

3   Case Study

This paper describes the development of an assurance case for a Tyre Noise Recognition (TNR) component that is used to improve multiple vehicle-level functions. The TNR makes use of microphones positioned within the wheel housing to measure road surface noise in order to determine, in real time, whether or not the road is dry. Here, dryness is defined as a road surface without any materials between tyre and road surface. This classification is, in turn, used as an additional source of information by chassis control and powertrain systems to determine the current surface traction and thereupon adapt control parameters accordingly, i.e. a *!dry* surface requires adaptations for a consistent traction. An overview of the architecture is depicted in Fig. 1.

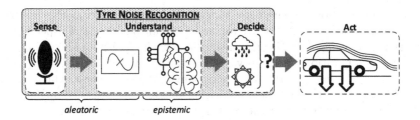

Fig. 1. TNR within system context and its sources of uncertainty.

In order to provide accurate information to the chassis control system, the TNR must process the audio signal with strict real-time requirements and be able to filter sampling anomalies caused by conditions such as the impact of loose gravel. Due to the runtime properties as well as the ability to process a wide range of signal patterns based on available data, a ML technique was chosen to implement the classification function of the TNR (cf. Sect. 5). Previous versions of the TNR were used to optimise chassis control performance. Through limits imposed within the vehicle-level function, the remaining safety concerns regarding the ML-based classification were low enough to assign only *Quality Management* (QM) requirements to the TNR after completing the hazard and risk analysis according to ISO 26262. However, in order to increase the functional benefits of the vehicle-level function through usage of TNR information, it was decided to evaluate the impact of reducing the limits imposed within the vehicle-level function. This in turn placed an increased safety load onto the TNR and hence led to the following *functional safety requirement* (FSR) allocated to the TNR:

– **FSR x:** The TNR shall not provide the result *dry* in case of a non-dry road surface (ASIL B).

The objective of the project was to develop an assurance case to argue that this level of integrity can be achieved for the TNR even though its output depends on a ML-based classification function. This includes ensuring that the hardware and software components were developed and verified according the *ASIL B* relevant guidelines of ISO 26262 to ensure the integrity of the execution with respect to hardware failures and software errors. In addition, SOTIF-like safety concerns regarding uncertainty in the domain understanding as well as accuracy of perception functions must also be considered when developing such novel ML-based perception systems. This work describes the underlying approach for ensuring a sufficient level of accuracy of the ML-based road surface classification across all target operational scenarios, providing a crucial building block for assuring the safety of ML-based systems for vehicle control systems.

4   Assurance Approach

ML as an implementation paradigm is increasingly used in automotive use cases where the characteristics of the environment can not be adequately specified for the purposes of an algorithmic implementation or where such an implementation may be too computationally intensive, as was the case in the TNR. This, however can come at the price of introducing uncertainty into the system, which in turn can manifest itself in the form of functional insufficiencies as defined by ISO/PAS 21448. These uncertainties manifest themselves in various components within the logical architecture of the system. Of particular interest for this work were the aleatoric uncertainty inherent in the environment in terms of the manifold factors that can impact the acoustic signal as well as the epistemic uncertainty introduced by the ML models themselves. Safety assurance must demonstrate that the system performance is able to satisfy the safety goals, despite these potential sources of inadequacies. Therefore, principles from the ISO/PAS 21448 were adapted to extend the safety lifecycle based on ISO 26262.

An additional factor increasing the difficulty of assuring safety is the issue of the *semantic gap* [4] by which the lack of a precise definition of the functional and performance requirements leads to an inadequate definition of safety requirements in relation to the intended or expected behaviour of the system. These considerations led to the identification of the following additional requirements on the safety lifecycle:

- A *domain analysis* as an extension of the *item definition* phase is required in order to include a thorough investigation of the operational domain and understanding of aspects of the environment that can lead to misclassifications. This phase led to an improved understanding of the system's safety requirements and the identification of a domain model, which in turn was used when reasoning about the completeness of training data and tests.
- The *design phase* refined these system-level requirements into technical safety requirements allocated to either primary functions or diagnostic and monitoring mechanisms. An analysis of potential *failure modes*, in terms of insufficiencies of the ML technique and model was required in order to identify performance improvements and diagnostic methods.
- Measures to *validate* the completeness of the specification and to determine whether a sufficient coverage of environmental conditions that lead to known insufficiencies has been reached and to minimise the residual risk of unknown triggering events.
- Due to the lack of specific guidance from the relevant safety standards, an *assurance case* approach [8] is required in order to reason about the adequacy of the safety approach. GSN [14] was applied in order to document, evaluate and argue the sufficiency of the safety measures within the project.

The phases of the assurance process (cf. Fig. 2) were implemented as an iterative process. For example, technical system design choices, such as the selection of sensor types, impact properties of the environment that must be analysed as part of the domain analysis. The discovery of unsupported assumptions in the assurance case may require a restriction of the functionality in order to ensure that the system safety requirements are fulfilled.

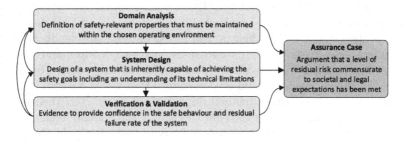

Fig. 2. Summary of the assurance process.

4.1 Domain Analysis

During the domain analysis phase, the open context environment was systematically investigated in order to understand factors influencing the sound profile and hence lead to unintended classifications. In order to focus on factors affecting safety, the following relationship between classified and actual prevailing road surface condition was established:

- True-Positive (TP)   Predicted *dry* while actually *dry*,
- True-Negative (TN)   Predicted *!dry* while actually *!dry*,
- False-Negative (FN)   Predicted *!dry* while actually *dry*,
- False-Positive (FP)   Predicted *dry* while actually *!dry*.

The misclassification *FN* only results in an overly conservative control strategy as higher traction is not actually needed but still activated, thereby not violating any safety goals. Hence, only the misclassification *FP*, which corresponds to *FSR x* (cf. Sect. 3), is safety-relevant.

Next, the concept of identifying *triggering events* as described in ISO/PAS 21448 was applied in order to develop an understanding of environmental conditions that could lead to a *FP* classification. This analysis was based on a thorough technical understanding of the sensing and signal processing principles involved as well as experience gained during the development and test of the previous QM-rated version of the TNR. The three main influences on the acoustic sensing that were identified from the environment are: tyres, road surface and the transmission medium of sound. These factors were then decomposed into their fundamental properties, e.g. tyres into rubber mixture, tyre pressure, tyre dimensions and others. The granularity and definition of each property has been selected according to physical realisability, for instance, tyre sizes only within actual produced dimensions. The resulting domain model consists of all feasible combinations of these properties and can be used to identify *known triggering events* describing known performance limitations of the systems [9]. The domain model can also be used to determine coverage criteria for test cases. However, even for this relatively simple application, the procedure created an unmanageably large amount of combinations. Too many, in fact, to be practically feasible. To reduce the amount of test cases, while still arguing coverage of the operational domain, each property and their individual impact was evaluated using expert knowledge. This assessment included considerations about safety with special attention to the physical sensing principle in detail, possible dependence between properties, as well as their overall significance for the classification. For instance, test cases regarding tyre dimensions only included min and max sizes and other combinations of parameters were considered irrelevant as no correlation between the parameters could be determined that would have an impact on the performance beyond the individual impact of the parameters themselves.

Nevertheless, uncertainty in the completeness of the domain model and adequacy of the abstractions required to reduce combinatorial explosion leads to the possibility of *unknown triggering events* and must still be accounted for in the assurance process. Therefore additional measures were defined in the V&V phase in order to validate the domain model.

4.2   System Design

To analyse the design of systems based on the TNR, the architecture presented in Fig. 1 was decomposed into the logical component groups: *Sense, Understand, Decide* and *Act*. This allowed for a clear separation of the concerns identified in the *Domain Analysis* and an analysis of each component's contribution to overall performance insufficiencies in the system.

The *sensing* part of the system includes the microphones inside the wheel arches and their task of measuring the sound waves. The sound waves are recoded within certain frequency boundaries and compressed for data transmission (cf. Fig. 1). Potential sources of aleatoric uncertainty are the lack of information, meaning the recorded frequency range does not cover the complete frequency spectrum sufficiently, measurement uncertainty, defining an imperfect measurement process by technical devices, and numerical approximations within the data compression algorithm. As all of these uncertainties can potentially lead to insufficient performance of the TNR, they have been addressed within the assurance case along with supporting evidence, e.g., mathematical analysis of data compression losses. The *understanding* portion of the logical architecture is accomplished by signal pre-processing and an ML-based classifier within the TNR (cf. Fig. 1). The classification exploits the fact that different road conditions are differentiable through acoustic properties. Potential causes of uncertainties are ambiguous sound patterns or epistemic uncertainty arising from the selected ML technique and model. A pessimistic decision strategy was used. In particular, the TNR will select the safer option *!dry* in case of conflicting predictions. Samples from multiple sensors are combined and aggregated over multiple sampling steps before providing a *dry* classification. The components corresponding to the *decide* and *act* function groups contain the chassis control logic and actuator components, respectively. According to the prediction the driving performance is optimised, for instance by adapting the suspension or spoiler.

4.3   Verification and Validation

Within the *verification and validation* (V&V) phase, performance requirements allocated to the system and its components were confirmed. In addition, assumptions regarding the performance potential of the design, as well as the environment operating conditions were confirmed in order to argue the safety of the system for its chosen context. This led to the identification of the following additional objectives within the V&V strategy, with respect to the QM version of the TNR function:

– **Confirmation of assumptions made during system design and safety assurance:** This included, for example, evaluating field data to assess whether the operating conditions matched the assumptions in the domain model (cf. Sect. 4.1) and confirming that pre-processing of the audio signal did not reduce the dimensionality of the input data in such a way that *dry* and *!dry* signals could be mapped to similar feature vectors. Other assumptions include the influence of signal noise (aleatoric uncertainty) on the performance of the classifier.

- **Evaluation of the function with regard to *known triggering events:*** These include combinations of environmental conditions discovered during the domain analysis as well as specific corner cases discovered during field testing.
- **Evaluation of the potential for *unknown triggering events:*** This objective includes confirms that the domain model covers a sufficient range of conditions that can impact the performance of the function and that all relevant usage scenarios have been considered.
- **Evaluation of the resilience of the function with regard to residual *unknown triggering events:*** This objective relates to the ability of the system to respond to signal patterns not considered in the domain model and for which either a valid response must nevertheless be given, or no value at all, resulting in a conservative action from the chassis control system.

A number of analyses, simulations and tests had previously been performed for the QM-rated version of the TNR. However, these measures were not necessarily aligned to the objectives described above, resulting in some gaps in the argumentation structure. Therefore, the following methods were identified in order to provide explicit evidence corresponding to the V&V objectives. In some cases, existing evidence could be aligned with the V&V objectives, in other cases, additional tests and associated documentation were required.

- **Analysis:** An understanding of the strengths and weaknesses of the chosen ML technique and model provided evidence for the inherent properties regarding robustness and generalisation. In addition, the prototypes generated by the algorithm (cf. Sect. 5) were amenable to examination by subject matter experts to confirm that they corresponded to known properties of the *dry* and *!dry* signals.
- **Simulation:** A simulation environment based on synthetic and recorded data was used for a focused verification of ML properties. Here, signal noise can also be simulated in order to verify the robustness of the classifier.
- **Structured testing:** The domain model was used to determine a set of test cases which cover all known properties which could influence the performance of the function. In addition, the test cases also included specific corner cases discovered during field tests and added to the regression test set.
- **Field tests:** Field tests, where the function was tested on real roads (both test track and public roads) were performed according to selected properties of the domain model (cf. Sect. 4.1). This allowed the coverage of conditions to be evaluated. Anomalies which could not be explained by the parameters of the domain model were used to iteratively refine the domain model.

4.4 Assurance Case

The objective of the assurance case was to develop a structured and convincing argument that the classifier fulfilled its technical requirements, in particular with respect to functional insufficiencies that could lead to *FP* identifications of dry road surface conditions. The assurance case was described using GSN and applied the principles from [8]. During the project it was primarily used as a means of communicating and evaluating the safety assurance approach within the team but was also developed with future external safety assessors in mind.

The top level structure of the assurance case is shown in Fig. 3. The assurance case focused on claims regarding a sufficient understanding of the domain and subsequent completeness of the technical safety requirements, the intrinsic performance potential of the chosen ML technique, the sufficiency of the training data to cover critical conditions of the domain, and the performance of the trained function itself. In addition, arguments were developed that the ML-based classifier was robust against changes in the operating environment as well as differences between the development and test environment and future deployment scenarios (e.g. different vehicle configurations).

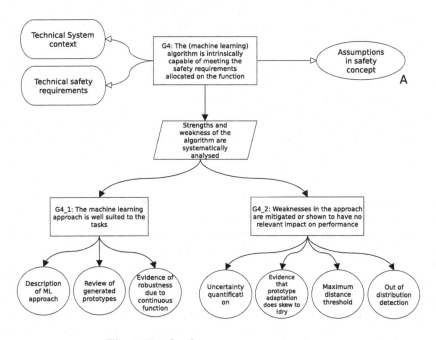

Fig. 3. Top-level assurance case structure.

5 Detailed Analysis of the Machine Learning Function

A detailed analysis of the applied ML-based classifier with respect to the technical safety requirements allocated to it and its general suitability regarding the target task was performed.

In the case of the TNR, *Adaptive Generalised Learning Vector Quantisation* (AGLVQ), an extension to GLVQ [13], was used. Figure 4 shows the operating principle of this algorithm.

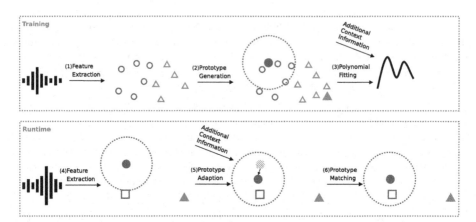

Fig. 4. Overview of the employed AGLVQ algorithm. In the training phase, meaningful features are extracted from the audio signal (1) that are subsequently used to generate prototypes (2) for both classes, *dry* (circles) and *!dry* (triangles), maximising the distance between them. A threshold for the *dry* prototype is defined, forming the decision space for this class. Additionally, a polynomial is fitted (3) that, using additional context information, adapts the *dry* prototype to the current situation. At runtime, features are again extracted from the audio signal (4), mapping the current sample (rectangle) to the feature space. After that, the learned polynomial is used to adapt (5) the *dry* prototype to the current situation. Finally, the current sample is matched to the prototypes (6) based on the Euclidean distance in the feature space. If the current sample is within the decision space of the *dry* prototype it is classified as such else a *!dry* road surface is assumed.

The use of AGLVQ had several advantages over other ML approaches from the perspective of safety assurance. The learned prototypes have been represented in the same form as the feature engineered audio signals and allowed the engineers to verify their plausibility. Additionally, in combination with the straightforward prototype matching used for runtime predictions and the interpretable adaption polynomial, it allowed for a detailed analysis of the decision space and uncovered potential error patterns. This also outlined another strength of this approach, the robustness to small input perturbations. Compared to, e.g., neural networks, there has been no feature subspace in which small changes of individual features could be amplified in a way that causes drastic and unexpected changes in the output. While not explicitly investigated, this may also significantly reduce the susceptibility to adversarial attacks. Due to this absence of discontinuities, the sensitivity to individual factors, e.g., tread depth, the burden of proof on the empirical tests was significantly lower compared to other discontinuous functions. Figure 5 shows an extract of the GSN regarding the choice of the ML technique. Here, the strengths of AGLVQ have been reflected in G4_1.

Several limitations of AGLVQ were also identified and used to derive additional *Technical Safety Requirements* (TSR). One such limitation was a low level of generalisation. As only a single prototype for the relevant class was generated there was a noticeable trade-off between safety and execution performance. A prototype adaption function mitigated this to some extent by incorporating additional knowledge about the context of the present situation. However, this was not sufficient for complex generalisations such as completely new types of road surfaces. Another limitation was that the prototype adaption in certain situations transformed the *dry* prototype slightly towards the decision space for non-dry road conditions, increasing the risk of incorrect classifications. Furthermore, the approach did not have an explicit way to quantify the uncertainty for the learning of, adapting, and matching to the prototype apart from the Euclidean distance, which does not fully account for the relation between the features. The known weaknesses and the evidence associated with the effectiveness of the counter-measures to these have been reflected in the GSN under G4_2. The analyses of these inherent weaknesses in the approach led to the proposal to develop self-assessment methods, specifically uncertainty quantification [5] and out-of-distribution detection to be applied at runtime.

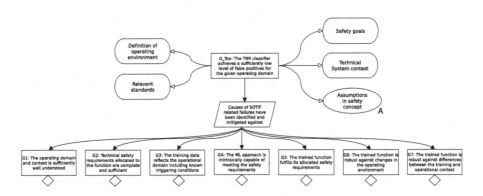

Fig. 5. Assurance case structure for choice of ML technique.

In order to demonstrate the performance of the trained function itself (sub-goal G5 of Fig. 5) performance metrics were defined and related to the TSRs. Since the TNR is a classification task, appropriate metrics were, among others, *accuracy, precision/recall* and *confusion matrices*. These helped to measure the overall performance and aided the investigation of error patterns. For instance, class-wise precision and recall allowed appropriate distance thresholds for the prototype matching to be identified and validated using the available test data. Additionally, the metrics helped with finding variances in the test data. The causes of these variances were iteratively analysed in more depth, either by gathering additional data or by qualitatively assessing the function with respect to the properties of the variance causing data. Lastly, the metrics were used to define acceptance criteria for the TSRs, e.g. that a certain class-wise accuracy on all validation datasets shall be achieved.

Based on a combination of the measures described above, the fundamental capability of the classifier for the target task was argued. Regarding the TNR, AGLVQ was found to be generally suitable, especially as the high degree of explainability allowed for a thorough analysis of the function and its behaviour. However, the known limitations and their consequences still left a burden of proof on the training data and the validation results, which were argued in sub-goals G3 and G5 of the assurance case.

6   Lessons Learned

The evaluation of the TNR with respect to its application for a safety-critical system (*ASIL B*) led to a number of lessons learned that could be applied in future projects as well as open questions that still remain to be resolved. The nature of the system level safety goals associated with the chassis control functions allowed for a safe state to be achieved if the road surface could be considered as *!dry*, thereby leading to a conservative traction control strategy. This allowed the function to be designed to indicate an invalid output in the case of ambiguous inputs as well as a skewing of the audio signals towards the *!dry* prototype if required.

The robustness and explainability of the approach helped with the in-depth analysis of the machine learning component. The ability to analyse the generated prototypes, their adaption to the current situation at runtime, and the respective decision space allowed the incorporation of expert knowledge in the quality assessment. In addition, the robustness due to the continuity of the function substantially facilitated the investigation of the influence of factors such as tread depth. This allowed for a significant reduction in the amount of in-field tests due to a reduction in the dimensions considered during coverage analysis.

Open questions nevertheless remain on the required level of granularity in the domain model used to evaluate the completeness of selected training data as well as quantitative test stopping criteria related to statistical performance metrics. Inevitably, an iterative approach to system development and assurance will be required (cf. Fig. 2) where field-based validation is required to confirm that sufficient detail in the domain model was achieved and that assumptions made during analysis, simulation and test were valid. These questions, however, are currently not addressed by existing safety standards such as ISO 26262, which assumes behaviour of software that can be evaluated through qualitative measures or ISO/PAS 21448 which requires a function-specific allocation of quantitative performance targets. The approach used within the project was to use qualitative arguments to argue the robustness of the ML function with respect to a broad range of operating conditions as defined by the domain model, whilst applying a range of measures (including extensive structured field tests) to confirm the assumptions behind the domain model. In addition, the TNR is embedded within a vehicle chassis control system, which in turn is developed and released according to a set of established development and homologation guidelines. Nevertheless, an external evaluation of the assurance approach by a qualified third party is recommended to examine the strength of the provided arguments.

7   Conclusion

The work described within this paper has demonstrated the feasibility of an assurance case for the application of ML for chassis control systems. The assurance approach made use of a systematic domain analysis to define properties of the environment relevant to the performance, dedicated measures in the system architecture to reduce the safety requirements on the ML function itself, the choice of an ML technique that enhanced robustness and explainability combined with a systematic validation plan to argue the absence of unknown triggering events. However, questions remain relating to the statistical level of performance that should be demonstrated by the ML algorithm. This type of evidence, would go above and beyond the forms of V&V proposed by the ISO 26262 standard for software but is required due to the inherent uncertainties when applying ML compared to conventional non data-driven algorithms.

The project highlighted the need for better industry-specific standards regarding the use of ML for safety-relevant functions, including outside of the domain of automated driving. These standards should include specific guidelines for determining coverage and selection criteria for training data, as well as for determining quantitative performance targets and testing criteria. These aspects would become even more relevant by alternative choices of ML technique, such as *Deep Neural Networks*, where qualitative arguments relating to the robustness and generalisation properties of the trained functions are more difficult to generate based on the complexity and opaqueness of the calculations involved. As such, any future standardisation should also include a differentiation of measures based on the intrinsic characteristics of the ML algorithms.

Acknowledgment. On side of Fraunhofer IKS, the research for developing the theoretical safety foundations utilisied in this work was partially funded by the Bavarian Ministry for Economic Affairs, Regional Development and Energy as part of a project to support the thematic development of the Institute for Cognitive Systems.

References

1. Bagschik, G., Reschka, A., Stolte, T., Maurer, M.: Identification of potential hazardous events for an unmanned protective vehicle. In: Proceedings of the IEEE Intelligent Vehicles Symposium (IV), Gothenburg, pp. 691–697 (2016)
2. Belinkov, Y.: On internal language representations in deep learning: an analysis of machine translation and speech recognition. Ph.D. thesis, Massachusetts Institute of Technology (MIT) (2018)

3. Burton, S., Gauerhof, L., Heinzemann, C.: Making the case for safety of machine learning in highly automated driving. In: Tonetta, S., Schoitsch, E., Bitsch, F. (eds.) SAFECOMP 2017. LNCS, vol. 10489, pp. 5–16. Springer, Cham (2017). https://doi.org/10.1007/978-3-319-66284-8_1

4. Burton, S., Habli, I., Lawton, T., McDermid, J., Morgan, P., Porter, Z.: Mind the gaps: assuring the safety of autonomous systems from an engineering, ethical, and legal perspective. Artif. Intell. **279**, 103201 (2020)

5. Fischer, L., Hammer, B., Wersing, H.: Efficient rejection strategies for prototype-based classification. Neurocomputing **169**, 334–342 (2015)

6. International Organization for Standardization: Road Vehicles: Functional Safety (ISO 26262) (2018)

7. International Organization for Standardization: Safety of the Intended Functionality - SOTIF (ISO/PAS 21448) (2019)

8. International Organization for Standardization: Systems and Software Engineering - ISO/IEC/IEEE 15026-1:2019 (2019)

9. Kurzidem, I., Saad, A., Schleiss, P.: A systematic approach to analyzing perception architectures in autonomous vehicles. In: Zeller, M., Höfig, K. (eds.) IMBSA 2020. LNCS, vol. 12297, pp. 149–162. Springer, Cham (2020). https://doi.org/10.1007/978-3-030-58920-2_10

10. Picardi, C., Hawkins, R., Paterson, C., Habli, I.: A pattern for arguing the assurance of machine learning in medical diagnosis systems. In: Romanovsky, A., Troubitsyna, E., Bitsch, F. (eds.) SAFECOMP 2019. LNCS, vol. 11698, pp. 165–179. Springer, Cham (2019). https://doi.org/10.1007/978-3-030-26601-1_12

11. Picardi, C., Paterson, C., Hawkins, R.D., Calinescu, R., Habli, I.: Assurance argument patterns and processes for machine learning in safety-related systems. In: Proceedings of the Workshop on Artificial Intelligence Safety (SafeAI), New York (2020)

12. Radlak, K., Szczepankiewicz, M., Jones, T., Serwa, P.: Organization of machine learning based product development as per ISO 26262 and ISO/PAS 21448. In: Proceedings of the 25th IEEE Pacific Rim International Symposium on Dependable Computing (PRDC), Perth, pp. 110–119 (2020)

13. Sato, A., Yamada, K.: Generalized learning vector quantization. In: Proceedings of the 8th International Conference on Neural Information Processing Systems (NIPS), pp. 423–429 (1995)

14. The Assurance Case Working Group (ACWG): Goal Structuring Notation - Community Standard. No. 2, Safety Critical Systems Club (SCSC) (2018)

15. Willers, O., Sudholt, S., Raafatnia, S., Abrecht, S.: Safety concerns and mitigation approaches regarding the use of deep learning in safety-critical perception tasks. In: Casimiro, A., Ortmeier, F., Schoitsch, E., Bitsch, F., Ferreira, P. (eds.) SAFECOMP 2020. LNCS, vol. 12235, pp. 336–350. Springer, Cham (2020). https://doi.org/10.1007/978-3-030-55583-2_25

16. Ye, H., Liang, L., Li, G.Y., Kim, J., Lu, L., Wu, M.: Machine learning for vehicular networks: recent advances and application examples. IEEE Veh. Technol. Mag. **13**(2), 94–101 (2018)

Safe Interaction of Automated Forklifts and Humans at Blind Corners in a Warehouse with Infrastructure Sensors

Christian Drabek[1]([✉]), Anna Kosmalska[1], Gereon Weiss[1], Tasuku Ishigooka[2], Satoshi Otsuka[2], and Mariko Mizuochi[3]

[1] Fraunhofer IKS, Munich, Germany
{christian.drabek,anna.kosmalska,gereon.weiss}@iks.fraunhofer.de
[2] Research and Development Group, Hitachi Ltd., Ibaraki, Japan
{tasuku.ishigoka.kc,satoshi.otsuka.hk}@hitachi.com
[3] Hitachi Europe GmbH, Schwaig, Germany
mariko.mizuochi@hitachi-eu.com

Abstract. Co-working and interaction of automated systems and humans in a warehouse is a significant challenge of progressing industrial systems' autonomy. Especially, blind corners pose a critical scenario, in which infrastructure-based sensors can provide more safety. The automation of vehicles is usually tied to an argument on improved safety. However, current standards still rely on the awareness of humans to avoid collisions, which is limited at corners with occlusion. Based on the examination of blind corner scenarios in a warehouse, we derive the relevant critical situations. We propose an architecture that uses infrastructure sensors to prevent human-robot collisions at blind corners with respect to automated forklifts. This includes a safety critical function using wireless communication, which sporadically might be unavailable or disturbed. Therefore, the proposed architecture is able to mitigate these faults and gracefully degrades performance if required. Within our extensive evaluation, we use a warehouse simulation to verify our approach and to estimate the impact on an automated forklift's performance.

Keywords: Driverless industrial trucks · Blind corners · Infrastructure sensors · Warehouse

1 Introduction

The progress of industrial automation leads to more and more integration of automated systems into today's industrial environments. Whereas previously the ideal procedure to follow was segregating automated machines from human workers, e.g., placing robots in dedicated safety cages, the co-working of humans and machines is an important factor for future competitiveness [23]. For instance, enabling personnel to be in the same area at the same time as automated guided vehicles (AGVs), provides ways to efficiently use the flexibility of humans and carrying power of machines [25].

© Springer Nature Switzerland AG 2021
I. Habli et al. (Eds.): SAFECOMP 2021, LNCS 12852, pp. 163–177, 2021.
https://doi.org/10.1007/978-3-030-83903-1_11

One example is a warehouse in which AGVs operate alongside workers. Industrial trucks are a major source of accidents within in-house transportation [6]. Additional guidelines for driverless industrial trucks [11], such as AGVs, reduce the inherent risk of human-machine-collisions. In order to not loose advantages of the automation by restricting operation to single areas, the pure local separation of human workers and AGVs cannot be maintained. In these scenarios, AGVs are equipped with safe perception capabilities, e.g., lidars or radars, and have to slow down or come to a complete stop, when obstacles in the surrounding are detected [26]. Additionally, guidelines for human workers should enforce compliance with the given safety rules. One major challenge in a modern warehouse, however, is the limited line of sight, for instance due to walls, shelves, or storage. In this case, corners become a potential point of risk. As AGVs cannot detect occluded objects at such intersecting paths, their application in this area is constrained, i.e. no operation or strongly reduced speed [1]. Human workers also cannot see around the corners and thus, are also in risk of provoking collisions [21]. Specific safety rules, like the rule of waiting at corners before stepping forward, could alleviate such risks, but are at risk of being ignored or overlooked during work. In general, safety mechanisms are more likely to be bypassed, if they are perceived to reduce efficiency [9]. In our work, we target to improve the safe interaction of humans and automated forklifts, which can be seen as a specific kind of AGV, by utilizing sensors of the infrastructure. Specifically, our approach addresses such corner situations, so-called 'blind corners', in warehouses operating automated forklifts. As main contributions, we analyze the problem of blind corners in detail and derive how present solutions manage the safety in such situations. Moreover, we present an architecture and new safety concepts exploiting infrastructure sensors for achieving a safe and efficient interaction of humans and automated forklifts at such critical situations. Our approach is evaluated by thorough simulations of blind corner warehouse situations and analyzing the adherence to safety goals and operation performance. With the results of this paper, we aim for showing how safety can be achieved, even in the presence of potentially unreliable wireless connections, by dynamically adjusting the forklift's performance with respect to the available perception information.

The remainder of this paper is structured as follows. Section 2 presents and formalizes the safety challenges of blind corners in common warehouses and introduces related work. In Sect. 3, we introduce our approach for infrastructure-based safe interaction of humans and automated forklifts at blind corners. Our extensive evaluation and results of our approach in varying warehouse scenarios are outlined in Sect. 4, before we conclude the paper in Sect. 5.

2   Blind Corners in Warehouses

2.1   Definition of Blind Corners

Walls or obstacles near the apex of a corner prevent any direct line of sight to crossing vehicles or humans. This situation is illustrated in Fig. 1. We define a *blind corner* as an intersection or turn that requires an ego vehicle to change its speed to avoid potential collisions, while line of sight is occluded by an obstacle. When approaching a blind corner, the required braking distance determines a safe speed limit until conflicting crossing objects can be excluded [35]. For road intersections, this can lead to a behavior similar to expert drivers [19]. However, compared to a road scenario, the safe deceleration of a forklift [24, 30] is lower and walls are often closer. To prevent the slowdown, information about the presence of a human (or other crossing vehicle) must be available much earlier [1].

For example, at a speed of $5\,\frac{m}{s}$, braking of a forklift needs to start at a distance d_{brake} of 3.5–6.5 m [30]. Additionally, the automated forklift travels $d_{process}$ while processing inputs. During processing, it needs to detect the intersection and if there is someone in the *conflict area*. Latter is defined by the time the forklift would take to pass the intersection and the passing human's speed:

$$d_{conflict} = v_{other}(d_{process} + d_{brake} + d_{inter} + d_{fl})/v_{fl}. \qquad (1)$$

If the conflict area and a margin for detection (d_{detect}) cannot be cleared, the forklift must decelerate to avoid a potential collision. The point for a decision based on line-of-sight [35] is close to the intersection, where the forklift already almost stops, as there is also less space to the occluding wall separating forklift and human. Further, this does not account yet for the whole length of the vehicle to pass. When using infrastructure sensors to avoid unnecessary slow downs of an approaching forklift, the sensors need at least a detection range with radius

$$R > d_{inter} + d_{conflict} + d_{detect}. \qquad (2)$$

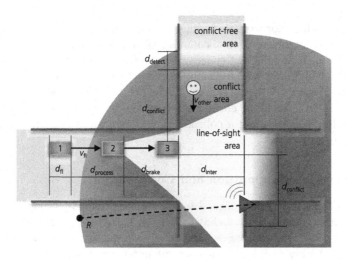

Fig. 1. An automated forklift approaching a blind corner.

2.2   Safety Standards for Driverless Industrial Trucks

As the autonomy of mobile machinery increases, also specific safety standards are being established. However, often there is a gap between requirements of these standards and state-of-the-art, which complicates more gradual paths to develop a system [34]. Within the European Union, laws such as the Machine Directive and national laws for protection of human safety are complemented by ISO and IEC standards that describe general design principles, cover aspects for a wide range of machinery or deal with particular machines [17]. The requirements for unmanned forklifts, AGVs and associated systems are defined by ISO 3691-4 [11].

Four kinds of access zones are defined by this standard. In the **operating zone**, a minimum clearance (i.e. 0.5 m wide) must be provided on both sides of the path and in the direction of travel. An **operating hazard zone**, where a person can be exposed to a hazard, requires audible or visual warnings and a low speed of $0.3 \frac{m}{s}$. Higher speeds like $1.2 \frac{m}{s}$ are only allowed under specific conditions. A **restricted zone** is a physically separated space, like a very narrow aisle, that may be entered only by authorized persons. Without personnel detection means, speed is limited to $0.3 \frac{m}{s}$. Access to enclosed space of a **confined zone** must be restricted to authorized personnel and is only allowed after the movement of trucks was stopped.

In this research, we examine the operation of automated forklifts within an *operating zone*. Therefore, they can go up to their rated speed and we assume a gap between forklift and the next wall of at least 0.5 m. Operation at rated speeds requires personnel detection mechanisms to be active. However, the standard requests a detection of persons in the direction of travel only, which is verified by testing if the forklift detects static cylinders representing legs of a standing worker or the body of a lying worker [11,13]. The result is that the forklift will not check for workers to its side and everywhere there is a blind corner. Safe interaction between an automated forklift and humans is only provided by the requirement of sufficient space around a forklift, which mitigates some of the more severe outcomes of a collision. Nevertheless, the main burden for avoiding collisions remains with the human. In future however, safety of a technical system is also envisioned to encompass freedom from danger [10]. Therefore, the responsibility of avoiding a collision should be moved from humans to the automated forklift.

2.3   Intersection Cooperation and Coordination

Blind corners are not a new hazard in warehouses. Human drivers of forklifts need to be instructed how to behave safely in such cases, e.g., slowing down, sounding the horn and looking around [4]. A simple flashing light, even when mounted in a highly visible location, might not be sufficient to prevent (near) collisions with a robot at a blind corner, as for example a reported case in October 1994 indicates [8]. Still, light spots or symbols projected into the direction of travel or around the forklift can improve awareness similar to beeper alarms for reversing [3].

Different technologies are currently researched that might enable drivers or autonomous machines in warehouses or on the road to see non-line-of-sight (NLOS) objects, e.g., around a blind corner. Selected examples are: the signal from surveillance cameras can be transformed to create virtual mirrors [12], moving shadows can be observed [20], radars can be used to detect moving objects around the corner [31,33] – possibly with the help of passive reflectors [29], and NLOS imaging can help to reconstruct hidden objects from multiply scattered light of laser sources [22]. Nevertheless, the computational power required to reach a sufficient performance and reliability level, so that safety-relevant decisions can rely on their measures, is just one reason why these technologies are still more a topic of the future.

By including support from infrastructure and infrastructure-based sensors, these problems can be avoided. For example, humans could be located in a warehouse using camera-data [14] or ultra wide band (UWB) technology [25,32] with a precision of at least 15 cm. Further, such a real-time locating system (RTLS) can also be used to predict the paths of workers [15]. Still, the safety integrity of such locating systems needs to be assessed. While at a higher cost, sensors similar to the safety equipment in automated forklifts could be installed at blind corners, to guarantee reliable detection of human workers. In addition, movement data can be collected and help in the creation of spaghetti charts to further analyze and improve safety [2].

Even if infrastructure can reliably identify workers, this information needs to be transferred to the automated forklift. Various methods to centrally coordinate vehicles and avoid collisions exist, e.g., [1,16,26,28]. However, the methods default to denying access to the intersection without connection or require a working connection. While reliability of connections can be improved by using multiple links [27], this also requires more resources. In the remainder of the paper, we detail how monitors for the infrastructure cooperation performance allow to dynamically adjust the forklift's actions to its available information.

3   Infrastructure-Cooperative Autonomous Control

In this section, we propose a novel architecture and compare multiple corresponding safety concepts as solutions for safe and efficient automated forklift operation in a warehouse, where human workers might be present. The description focuses on interactions at blind corner. Safe and efficient operation of autonomous systems in cooperation with humans is usually handled by reducing the machine's speed when humans approach [23]. Blind corners require support from infrastructure to detect human workers efficiently [1]. This includes a safety critical function using wireless communication, which sporadically might be unavailable or disturbed. The architecture which we propose is able to mitigate these faults and automatically adjusts the performance if required. In the next section, we provide a quantitative evaluation of the concept's influences on safety and efficiency, which is intended to help selecting the appropriate safety concept according to different conditions and requirements of warehouse operation.

3.1   Infrastructure-Cooperative Autonomous Control Architecture

This subsection presents our architecture to achieve safe and efficient operation of automated forklifts in a warehouse where human workers can be in proximity. The architecture includes the core AGV tasks [5] and utilizes infrastructure sensors and systems to monitor, predict and estimate the risk of hazardous situations in the warehouse. Additionally, the cooperation of infrastructure and forklift is continuously monitored on both sides to adjust the performance, e.g., the speed of the forklift, if required for safe operation. Figure 2 shows an overview of the proposed infrastructure-cooperative architecture for automated forklifts.

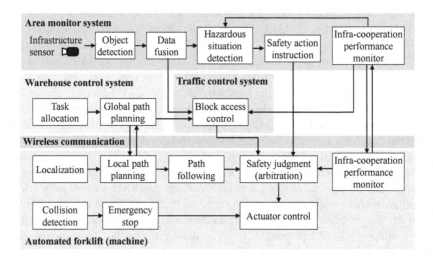

Fig. 2. Infrastructure-cooperative autonomous control architecture

The **warehouse control system** aims to maximize overall operational efficiency. *Task allocation* assigns tasks to each forklift (and worker) considering overall efficiency, where *global path planning* determines an optimal route for each forklift and task.

The **traffic control system** coordinates the (automated) movement in the warehouse. For example, no collision of machines will occur if only one machine may enter a certain area at the same time. Therefore, *block access control* manages the permissions of machines to enter blocks along their planned paths based on available information, e.g., positions and paths.

The aim of the **area monitor system** is to avoid the collision risk that cannot be prevented by the traffic control system. The system monitors the existence and movement of machines and workers in the warehouse using available infrastructure sensors for *object detection* and *data fusion*. *Hazardous situation detection* recognizes defined safety risks or deviations from rules. The system determines and issues *safety action instructions*. It can request a connected forklift to follow them immediately if a safety risk is observed or prepare the case of a missing connection.

The **automated forklift** is provided with various functions to ensure safe operation of the machine. Based on the position of the forklift identified by *localization*, its trajectory is determined by *local path planning* and *path following* based on the designated route provided by the warehouse control system. *Safety judgment (arbitration)* validates the trajectory and determines a suitable speed that ensures safe operation based on received permissions from traffic control, safety action instructions from area monitor and the reported status of *infra-cooperation performance monitor*. The *collision detection* and *emergency stop* functions implement the personnel detection mechanisms required by current standards, e.g., the ISO 3691-4 [11], using the machine's own sensors.

In brief, the proposed architecture ensures safety in three ways: block permission, area monitor and emergency stop. This structure enables collision avoidance in advance and reduces unnecessary deceleration and stoppage of the automated forklift. The operational efficiency can, thus, be expected to be improved. However, new potential hazards or failures are introduced when the safety critical function uses information from infrastructure systems.

3.2 Infrastructure-Cooperative Autonomous Control Hazards

This subsection examines the potential hazards of including infrastructure information in a safety critical function, like collision avoidance at blind corners from a functional architecture perspective. Therefore, the fault-tree shown in Fig. 3 analyzes the functional interactions of the subsystems and does not consider any hardware or implementation faults.

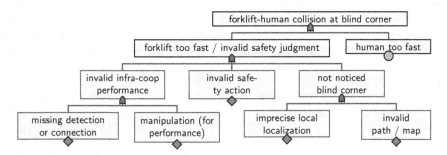

Fig. 3. Fault-tree of forklift-human collision for the functional architecture.

The forklift could be too fast, e.g. by failing to slow down at a blind corner due to making an invalid safety judgment. This can happen if the forklift does not notice the blind corner situation because of an incorrect position; either, because of insufficient localization, or by using an invalid map, which might be outdated or incompatible with the path received from warehouse management. Further, it could be too fast by an invalid safety action or by an invalid monitoring of infra-coop performance. The architecture is designed to handle missing detections and connections gracefully by monitoring. However, manipulation poses an inherent potential safety risk. Manipulation of safety mechanisms is often motivated by improved performance [9]. For example, a human working near an intersection could be spoofing the sensor to prevent forklifts from slowing down. Such cases can be handled safer, if the safety concept allows integrating performance concerns. Besides the forklift, also a faster than anticipated human worker could reach the intersection early, even though he was not in the designated conflict area when the automated forklift had to make its decision.

Besides these functional faults, we identified **the dependability of detection using infrastructure sensors** and the **dependability of wireless communication** between systems as main sources of this hazard.

Sensors and algorithms used for detection in the infrastructure system need to comply with necessary safety levels, e.g., they must be either able to reliably detect the presence of humans in the required range or recognize their inability to do so. The former can be achieved, for example, by using sensors similar to those used on the vehicle to detect humans.

Wireless communication, in the fault-tree contained within *invalid infra-coop performance*, cannot be as easily guaranteed to always work. A loss of connection cannot be avoided perfectly, even with efforts to improve the reliability of wireless communication. Since the proposed architecture requires continual exchange of information between the automated forklifts and the infrastructure systems, a mechanism to continue safe and efficient operation even with a failure in the communication is indispensable.

As a mitigation for this weakness, we introduce a monitoring and recovery mechanism. The infrastructure cooperative performance monitor in Fig. 2 monitors the condition of communication between automated forklift and area monitor. If communication fails longer than a predetermined interval, the forklift can switch to a mitigation mode using its own sensors only [7]. Safety judgment ensures safety for this *conditional selection of personnel detection means protected zones* [11]. However, this might interfere with decisions made by traffic control. As a mitigation measure on the infrastructure side, for example, it is conceivable to regard the loss of communication as a safety risk and notify each function of the situation. The block access control and hazardous situation detection can adapt instructions to other forklifts if necessary. Details of the forklift's local mode will be described together with safety concept in the next subsection.

3.3 Safety Concepts for Safe Interaction of Automated Forklifts and Human Workers at Blind Corners

This subsection presents three safety concepts for safe interaction of the automated forklifts and human workers at blind corners, based on the proposed architecture. Movement of human workers can only be controlled by signals and operational rules. However, they can be ignored or violated for various reasons [3,4,8,10], intentionally or unintentionally. On the other hand, a safety concept suitable for individual warehouses is not always the same.

This research considers the following alternatives as *safety concept (SC)* for safe interaction at blind corners:

- SC0: Stop only for humans detected in direction of travel
- SCA: Decelerate at blind corners and prioritize forklift
- SCB: Utilize infrastructure and prioritize forklift
- SCC: Utilize infrastructure and prioritize person

SC0 uses only the minimum personnel detection mechanism required by ISO 3691-4 [11] outside of confined zones. All other responsibility to avoid collisions at blind corners remains with the worker.

SCA is similar to a conventional operation with human-driven forklifts [4,30] without using infrastructure systems. As operational rule, *persons should pause at intersections, check if a forklift is approaching, and wait for the forklift to pass; and the forklift is allowed to continue slowly.* Even though priority is given to forklifts, they must pass intersections while paying attention to the presence of human workers that could violate the safety rule. As it is difficult for the forklift to detect persons due to blind corners, intersection areas are treated like operating hazard zones and the forklift's speed is limited accordingly. The deceleration helps to avoid collisions, but may unnecessarily hinder efficiency.

SCB utilizes the proposed infrastructure cooperative architecture and prioritizes passage of forklifts. The operational rule for the worker is identical to SCA, but an *automated forklift may continue at normal speed, if there is no worker in the conflict area.* Presence of workers in the *conflict area* is monitored by infrastructure sensors and causes a safety action instruction to slow down for the automated forklift. In addition, operation of the forklift is switched to a *local mode* similar to SCA if a missing safety action instruction or a failed communication is detected, e.g. by using a heartbeat with a rolling counter for the instructions. This safety concept is expected to improve operational efficiency by limiting the situations that require significant deceleration of the forklift. On the other hand, the risk remains that persons violating the rule and stepping into intersections may cause collisions with the slow forklift.

SCC utilizes the proposed infrastructure cooperative architecture and prioritizes passage of human workers. The operational rule is set as follows: *Forklifts check the existence of workers in conflict areas and wait for them to pass; and human workers do not have to stop and can pass freely.* Like in SCB, infrastructure sensors monitor the presence of workers and the forklift can pass at a normal speed in their absence. Also, a *local mode* similar to SCA is used if a missing safety action instruction or a failed communication is detected. The difference between SCC and SCB is the triggered safety action, if a worker is present. In SCC, the forklift is instructed to decelerate, stop and wait for the worker to pass. The forklift can resume passing only after all humans left the conflict area. As priority is given to humans, they cannot violate a rule preferring forklifts. However, unnecessary waiting times for forklifts may occur, especially if a human worker stays near an intersection.

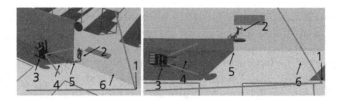

Fig. 4. Overview of the blind corner scenarios used in the simulation: (1) infrastructure camera, (2) human worker, (3) forklift, (4) forklift cameras field of view, (5) blind corner, (6) infrastructure camera field of view.

4   Evaluation

The safety concepts are evaluated in a robotics simulation using Webots [18]. For this, the architecture shown in Fig. 2 was implemented and several scenarios in the warehouse setting have been examined at the intersection shown in Fig. 4. In the simulation, cameras with correct object recognition are used as assumed dependable sensors. A fixed infrastructure sensor covers the *conflict area* of the intersection. It recognizes forklifts, human workers and their positions. If a human is detected in the conflict area, the selected safety action is forwarded to the approaching forklift. Block control and warehouse control system have been replaced with stubs that instruct the forklift to pass the intersection.

The automated forklift is equipped with two sensors, covering the area in front of the forklift and partially on the sides, simulating the personnel detection mechanism required by ISO 3691-4 [11]. In each simulation run, the forklift has to travel 35 m starting 25 m before the intersection. The simulation assumes a rated speed of 5 $\frac{m}{s}$ and a value of 3 $\frac{m}{s^2}$ for brake and acceleration. For simplicity, the simulated human worker follows a straight path across the intersection and will either pass it, wait before the intersection (but already in conflict area) or not enter the conflict area. To cover all cases where a human and a slow or fast moving forklift would arrive at the intersection at the same time and collide without any further action, 45 different initial distances of the human to the intersection were selected. The resulting scenarios range from the human worker crossing the intersection before the automated forklift to a slow forklift passing before the human can reach the blind corner.

The safety concepts *SC0*, *SCA*, *SCB* and *SCC* have been implemented in the Webots simulation, allowing the comparison of their performance and safety. A shorter **average completion time** indicates a better performance of the system. We only take successfully completed runs into account, i.e., runs with collisions or timeouts (30 s) are excluded. The **number of collisions** indicates the safety of a system – ideally there should be no collisions. Front and side collisions are considered as simplified indication of an accident's severity. Side collisions are less severe and result from humans walking into a visible forklift.

Table 1. Simulation results: average time of successful completion of task ($t_{avg}[s]$) and number of simulations that finished successfully (F), ended due to timeout (T), ended in front collision (fC) or side collision (sC). Zero values have been omitted for clarity.

Human	Metric	Communication status							
		Off		On		On→Off		Off→On	
		SC0	SCA/B/C	SCB	SCC	SCB	SCC	SCB	SCC
Human	Metric	0	1	2	3	4	5	6	7
Passes intersection	t_{avg}	7.3	12.6	8.7	9.1	12.2	12.2	8.8	9.2
	F	29	37	45	45	37	37	45	45
	fC	13	3			3	3		
	sC	3	5			5	5		
Waits in conflict area	t_{avg}	7.3	11.7	9.9	7.3	11.4	11.4	10	7.4
	F	45	45	45	18	45	45	45	17
	T				27				28
Not present	t_{avg}	7.3	11.7	7.3	7.3	10.9	10.9	7.3	7.3
	F	45	45	45	45	45	45	45	45

For each safety concept, all possible combinations of communication with infrastructure status and human behavior were taken into account. Table 1 summarizes the average completion times and the different results for each combination. While there is only a collision risk, if the human crosses the intersection, good performance in the other cases is expected to improve acceptance of a selected safety concept. If communication is off, forklifts ignore all messages from infrastructure. Switching of communication status was timed to impact right before the forklift has to make its decision whether to brake or not. Since the scenarios cover all possible encounters that can be achieved by changing the start positions, we assume that the following observations apply in general to the interaction of an automated forklift and a human worker at a blind corner:

SC0 provides very good performance when there is no risk of collision. The average time of performing a task is 7.3 s, which is the best achievable result. However, if the human does not prioritize the forklift and walks into the intersection, more than every third simulation run ended in a collision. This result underlines the motivation that intersections with blind corners require special attention to ensure safety for human-machine cooperation.

SCA results in less front collisions than *SC0*, by slowing down near the blind corner. This concept increases safety by giving the forklift more time to detect a person and stop – only 7 out of 45 runs ended in collisions, including 3 severe accidents. However, it increases the average completion time to almost 12 s, as deceleration is performed regardless of a nearby human.

SCB provides very good performance when there is no human in the conflict area and the forklift receives permission to pass the blind corner at maximum rated speed. In these scenarios, this also avoided collisions that could only have

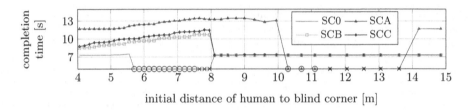

initial distance of human to blind corner [m]

Fig. 5. Forklift completion times for passing humans based on the human's initial position. Circles and crosses mark runs resulting in front or side collisions respectively.

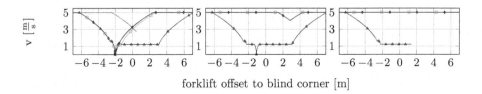

forklift offset to blind corner [m]

Fig. 6. Speed curves of the forklift for human starting at 6,9 and 12 m distance.

happened, if the forklift had slowed down. When needed, slowing down provided sufficient delay for the forklift to detect the human and stop.

In the observed cases, SCC provides no significant improvement over SCB. However, SCC can improve safety in more complex scenarios, e.g., if the human is not moving with constant speed or if multiple humans are present. On the other hand, if a human remains in the conflict area, the forklift will also wait indefinitely, unless an additional override mechanism is implemented. For both, SCB and SCC, the forklift gracefully degrades to SCA's safety and performance level, if it receives no information from infrastructure.

The impact of the human's initial position on completion times is shown in Fig. 5. Two areas of potential collisions can be identified in the diagram. If the human is already close to the intersection, he could collide with a quick moving forklift ($SC0$). A slow moving forklift (SCA) could collide with a human further away. The dynamic decision being made in SCB and SCC allows the forklift to avoid these situations by either slowing down for the human to cross (left) or passing the intersection quickly and safely before the worker can reach it (right). For the human starting at distances of 6, 9 and 12 m to the intersection, speed curves of the forklift are shown in Fig. 6 from left to right. In the left diagram, forklifts decelerate at d_{brake} except for $SC0$, for which the emergency brake could not prevent a collision. In the other diagrams, the forklift decelerates only for SCA: In the middle, the forklift avoids a collision by stopping, while an inattentive human runs into the forklift's side on the right.

5   Conclusion and Outlook

This paper examines the interactions of automated forklifts and humans at blind corners in a warehouse. We introduce an architecture that includes infrastructure sensors to increase the safety in these situations while having minimal impact on efficiency. We present and compare safety concepts related to this architecture that each address different needs. Clearly, relying only on the forklift's own sensors either poses a high risk for human workers, if the forklift does not slow down at intersections (*SC0*) or suffers a severe performance penalty (*SCA*). Using information from infrastructure sensors (*SCB*), the decision to slow down can be made dynamically, which reduces the impact on performance, even if the connection is not always available. Still, a small risk remains if a slowly approaching automated forklift is ignored. However, instructing forklifts to unconditionally stop for humans (*SCC*) will lead to unnecessary waiting times.

In the future, a human worker's behavior might be inferred automatically, when it is possible to have more and reliable information, e.g. by an improved prediction that can recognize the human's intention and awareness. In the meantime, such systems could include means for workers to actively yield to forklifts.

Acknowledgment. The research leading to these results has partially received funding from the Bavarian Ministry of Economic Affairs, Regional Development and Energy as Fraunhofer High Performance Center Secure Intelligent Systems.

References

1. Boehning, M.: Improving safety and efficiency of AGVs at warehouse black spots. In: IEEE ICCP, pp. 245–249, September 2014. https://doi.org/10.1109/ICCP.2014. 6937004
2. Cantini, A., De Carlo, F., Tucci, M.: Towards forklift safety in a warehouse: an approach based on the automatic analysis of resource flows. Sustainability **12**(21), 8949 (2020). https://doi.org/10.3390/su12218949
3. Cao, L., Depner, T., Borstell, H., Richter, K.: Discussions on sensor-based assistance systems for forklifts. In: Smart SysTech, pp. 1–8, June 2019
4. Cohen, H.H., Jensen, R.C.: Measuring the effectiveness of an industrial lift truck safety training program. J. Saf. Res. **15**(3), 125–135 (1984). https://doi.org/10. 1016/0022-4375(84)90023-9
5. De Ryck, M., Versteyhe, M., Debrouwere, F.: Automated guided vehicle systems, state-of-the-art control algorithms and techniques. J. Manuf. Syst. **54**, 152–173 (2020). https://doi.org/10.1016/j.jmsy.2019.12.002
6. Arbeitsunfallgeschehen 2019: Statistik 21537, DGUV, September 2020. https:// publikationen.dguv.de/widgets/pdf/download/article/3893
7. Drabek, C., et al.: Dependable and efficient cloud-based safety-critical applications by example of automated valet parking. In: Martins, A.L., Ferreira, J.C., Kocian, A., Costa, V. (eds.) INTSYS 2020. LNICST, vol. 364, pp. 90–109. Springer, Cham (2021). https://doi.org/10.1007/978-3-030-71454-3_6
8. Everett, H.R., Gage, D.W., Gilbreath, G.A., Laird, R.T., Smurlo, R.P.: Real-world issues in warehouse navigation. In: Mobile Robots IX, vol. 2352, pp. 249–259. SPIE, Boston, January 1995. https://doi.org/10.1117/12.198975

9. Manipulation von Schutzeinrichtungen - Verhindern, Erschweren, Erkennen. Fachbereich AKTUELL FB HM-022, FB HM DGUV, July 2016
10. Safety in the future: Whitepaper, IEC, Geneva, Switzerland, November 2020. https://go.iec.ch/wpsif
11. Industrial trucks: Safety requirements and verification: Part 4: Driverless industrial trucks and their systems. International Standard ISO 3691-4:2020(E) (2020)
12. Kojima, K., Sato, A., Taya, F., Kameda, Y., Ohta, Y.: NaviView: visual assistance by virtual mirrors at blind intersection. In: ITSC, pp. 592–597, September 2005. https://doi.org/10.1109/ITSC.2005.1520120
13. Korte, D.: Sicherheitsbezogenes Sensorsystem für fahrerlose Transportfahrzeuge. Logist. J. **2020**(12) (2020). https://doi.org/10.2195/LJ_PROC_KORTE_DE_202012_01
14. Košnar, K., Ecorchard, G., Přeučil, L.: Localization of humans in warehouse based on rack detection. In: ECMR, pp. 1–6, September 2019. https://doi.org/10.1109/ECMR.2019.8870913
15. Löcklin, A., Ruppert, T., Jakab, L., Libert, R., Jazdi, N., Weyrich, M.: Trajectory prediction of humans in factories and warehouses with real-time locating systems. In: IEEE ETFA, vol. 1, pp. 1317–1320 (2020). https://doi.org/10.1109/ETFA46521.2020.9211913
16. Lombard, A., Perronnet, F., Abbas-Turki, A., El Moudni, A.: Decentralized management of intersections of automated guided vehicles. IFAC-PapersOnLine **49**(12), 497–502 (2016). https://doi.org/10.1016/j.ifacol.2016.07.669
17. Markis, A., Papa, M., Kaselautzke, D., Rathmair, M., Sattinger, V., Brandstotter, M.: Safety of mobile robot systems in industrial applications. In: Proceedings of the ARW & OAGM Workshop, Steyr, Austria, pp. 26–31 (2019). https://doi.org/10.3217/978-3-85125-663-5-04
18. Michel, O.: Cyberbotics Ltd. Webots™: professional mobile robot simulation. J. Adv. Robot. Syst. **1**(1), 39–42 (2004). https://doi.org/10.5772/5618
19. Morales, Y., Yoshihara, Y., Akai, N., Takeuchi, E., Ninomiya, Y.: Proactive driving modeling in blind intersections based on expert driver data. In: IEEE IV, Los Angeles, CA, USA, pp. 901–907, June 2017. https://doi.org/10.1109/IVS.2017.7995830
20. Naser, F., et al.: ShadowCam: real-time detection of moving obstacles behind a corner for autonomous vehicles. In: ITSC, Maui, HI, USA, pp. 560–567 (2018). https://doi.org/10.1109/ITSC.2018.8569569
21. Okamoto, T., Yamada, Y.: Study of conditions for safe and efficient traffic in an indoor blind corner-based decision model with consideration for tactics and information uncertainty. In: 2012 IEEE RO-MAN, pp. 682–688, September 2012. https://doi.org/10.1109/ROMAN.2012.6343830
22. O'Toole, M., Lindell, D.B., Wetzstein, G.: Confocal non-line-of-sight imaging based on the light-cone transform. Nature **555**(7696), 338–341 (2018). https://doi.org/10.1038/nature25489
23. Platbrood, F., Görnemann, O.: Safe Robotics – die Sicherheit in kollaborativen Robotersystemen. Whitepaper 8020620, SICK AG, June 2018
24. Railsback, B.T., Ziernicki, R.M.: Stand-up forklift acceleration. In: ASME IMECE, pp. 421–424. ASMEDC, Vancouver, November 2010. https://doi.org/10.1115/IMECE2010-38940
25. Rey, R., Corzetto, M., Cobano, J.A., Merino, L., Caballero, F.: Human-robot coworking system for warehouse automation. In: IEEE ETFA, pp. 578–585 (2019). https://doi.org/10.1109/ETFA.2019.8869178

26. Sabattini, L., et al.: The PAN-robots project: advanced automated guided vehicle systems for industrial logistics. IEEE Robot. Autom. Mag. **25**(1), 55–64 (2018). https://doi.org/10.1109/MRA.2017.2700325

27. Scheuvens, L., Hößler, T., Barreto, A.N., Fettweis, G.P.: Wireless control communications co-design via application-adaptive resource management. In: 2019 IEEE 2nd 5G World Forum (5GWF), pp. 298–303, September 2019

28. Shirazi, M.S., Morris, B.T.: Looking at intersections: a survey of intersection monitoring, behavior and safety analysis of recent studies. IEEE Trans. Intell. Transp. Syst. **18**(1), 4–24 (2017). https://doi.org/10.1109/TITS.2016.2568920

29. Solomitckii, D., Barneto, C.B., Turunen, M., Allén, M., Koucheryavy, Y., Valkama, M.: Millimeter-wave automotive radar scheme with passive reflector for blind corner conditions. In: EuCAP, Copenhagen, Denmark, pp. 1–5, March 2020. https://doi.org/10.23919/EuCAP48036.2020.9135926

30. Forklift safety - reducing the risks. Technical report, State of Queensland (2019). https://www.worksafe.qld.gov.au/__data/assets/pdf_file/0021/21459/forklift-safety-reducing-risks-guide.pdf

31. Sume, A., et al.: Radar detection of moving targets behind corners. IEEE Trans. Geosci. Remote Sens. **49**(6), 2259–2267 (2011). https://doi.org/10.1109/TGRS.2010.2096471

32. Sun, E., Ma, R.: The UWB based forklift trucks indoor positioning and safety management system. In: IEEE IAEAC, pp. 86–90, March 2017. https://doi.org/10.1109/IAEAC.2017.8053982

33. Thai, K., et al.: Around-the-corner radar: detection and localization of a target in non-line of sight. In: IEEE RadarConf, Seattle, WA, USA, pp. 0842–0847, May 2017. https://doi.org/10.1109/RADAR.2017.7944320

34. Tiusanen, R., Malm, T., Ronkainen, A.: An overview of current safety requirements for autonomous machines – review of standards. Open Eng. **10**(1) (2020). https://doi.org/10.1515/eng-2020-0074

35. Yoshihara, Y., Morales, Y., Akai, N., Takeuchi, E., Ninomiya, Y.: Autonomous predictive driving for blind intersections. In: IEEE/RSJ IROS (2017). https://doi.org/10.1109/IROS.2017.8206185

Machine Learning-Based Fault Injection for Hazard Analysis and Risk Assessment

Bentley James Oakes[1,2], Mehrdad Moradi[1,2(✉)], Simon Van Mierlo[1,2],
Hans Vangheluwe[1,2], and Joachim Denil[1,2]

[1] University of Antwerp, Antwerp, Belgium
[2] Flanders Make vzw, Lommel, Belgium
{bentley.oakes,mehrdad.moradi,simon.vanmierlo,hans.vangheluwe,
joachim.denil}@uantwerpen.be

Abstract. Current automotive standards such as ISO 26262 require
Hazard Analysis and Risk Assessment (HARA) on possible hazards and
consequences of safety-critical components. This work attempts to ease
this labour-intensive process by using machine learning-based fault injection to discover representative hazardous situations. Using a Simulation-
Aided Hazard Analysis and Risk Assessment (SAHARA) methodology,
a visualisation and suggested hazard classification is then presented for
the safety engineer. We demonstrate this SAHARA methodology using
machine learning-based fault injection on a safety-critical use case of an
adaptive cruise control system, to show that our approach can discover,
visualise, and classify hazardous situations in a (semi-)automated manner in around twenty minutes.

Keywords: Hazard analysis · Risk assessment · Verification · Fault
injection · Reinforcement learning · Signal temporal logic

1 Introduction

Automotive systems are complex *cyber-physical systems* with ever-tightening
safety and production efficiency requirements. These qualities must be ensured
even as automotive software contains lines of code numbering in the tens of
millions [6] to support many modern features, including autonomous operation.

The well-known ISO 26262 standard mandates manufacturers to perform
safety and hazard analysis of their vehicles [11]. In particular, *hazards* and *faults*
must be shown to be adequately considered and handled by the manufacturer
in the form of supporting evidence cases. ISO 26262 defines one outcome of
this analysis as the Automotive Safety Integrity Level (ASIL), which denotes
the risk of a hazard and, therefore, the level of risk reduction required to be

This work was partly funded by Flanders Make vzw, the strategic research centre for the
Flemish manufacturing industry; and by the aSET project (grant no. HBC.2017.0389)
of the Flanders Innovation and Entrepreneurship agency (VLAIO).

I. Habli et al. (Eds.): SAFECOMP 2021, LNCS 12852, pp. 178–192, 2021.
https://doi.org/10.1007/978-3-030-83903-1_12

implemented by automotive components. Performing this *Hazard Analysis and Risk Assessment* (HARA) is manual work that requires hours of discussions between safety engineers [16]. Automating aspects of this analysis, therefore, greatly reduces the time taken to understand hazardous situations.

A *Simulation-Aided Hazard Analysis and Risk Assessment* (SAHARA) approach can utilise intelligent Fault Injection (FI) on vehicle components, which are then simulated for the visualisation and classification of hazardous situations [21]. Further described in Sect. 3, this SAHARA methodology[1] involves models of the components under study, information on how to inject faults, the safety-critical scenarios of interest, and a process to suggest a hazard classification for the simulation result. This (semi-) automated methodology provides safety engineers with representative visualisations and safety classifications of a system's faulty behaviour in various hazardous situations.

However, the SAHARA methodology defined in [21] presents few details on the FI procedure and implementation. For example, selecting the injection site and optimising the fault parameters is not discussed, and no indication of the performance of the FI procedure is given.

Our research focuses on intelligent FI, which employs a Reinforcement Learning (RL) algorithm to discover fault parameters [18]. RL is a large category of Machine Learning (ML) algorithms learning from the environment by interacting dynamically with it [20]. RL is used to automatically identify the parameters for critical faults that should be injected into the component under test to provoke increasingly more hazardous behaviour. In this work, we place our FI approach within the SAHARA methodology and inject hazardous faults. The CARLA open-source simulator for automotive research [8] then produces a visualisation of the resulting system behaviour, and an ASIL is suggested using temporal logic on the simulation traces.

This paper's contributions are therefore: (i) detailing how the ML-based FI process uses the available data within the SAHARA methodology to automatically produce hazardous situations, (ii) providing an example of the FI and SAHARA processes on a use case, including an indication of the approach performance, and (iii) a discussion of the benefits and drawbacks of placing ML-based FI within the SAHARA process.

Section 2 introduces the adaptive cruise control example. Section 3 describes FI within the SAHARA methodology, while Sect. 4 an indication of performance. Section 5 discusses the approach, while related literature is presented in Sect. 6 and Sect. 7 concludes the paper and describes future work.

2 Adaptive Cruise Control

This section introduces the Adaptive Cruise Control (ACC) system under study and the potential hazards that may arise from faults in the component.

---

[1] Other approaches such as [13,14] refer to a SAHARA approach. This paper uses SAHARA solely to refer to the methodology of [21].

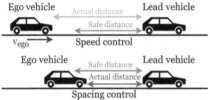
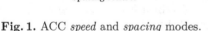

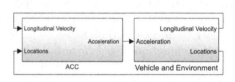

Fig. 1. ACC *speed* and *spacing* modes.   **Fig. 2.** ACC model in Simulink®.

The purpose of the ACC is to regulate the speed of the *ego vehicle* (the vehicle of concern) to ensure that it does not approach the rear of any *lead vehicles* too closely. The user of the ACC defines a *preferred speed* and *safe distance*, which is compared to the *relative distance* between the ego and lead vehicles. Figure 1 presents the two modes of the ACC. The first mode of the ACC is the *speed control* mode; the ACC directs the ego vehicle to increase the vehicle's speed to the preferred speed, potentially decreasing the relative distance. The second *spacing control* mode occurs when the ego vehicle is within safe distance to the lead vehicle. Here, the ACC directs the ego vehicle to reduce its speed such that the safe distance is maintained.

The Simulink® model used for this work is shown in Fig. 2, where the *ACC* is modelled in the left-hand block and the *ego vehicle dynamics and environment* are modelled in the right-hand block[2]. In this work, the automotive simulator CARLA [8] will replace the vehicle dynamics and environment block.

As the ACC component can control the vehicle's acceleration and the resulting distance from other vehicles, the ACC is a safety-critical component. Even a minor fault could violate the safety requirement that the relative distances between vehicles is greater than a set safe distance. In a more hazardous situation, a major fault in the ACC could lead to an unintended acceleration into the rear of the lead vehicle, potentially resulting in severe injuries or death. Therefore, this ACC component must be intensively examined in a structured manner to determine possible hazards and their consequences.

In Sect. 3.3, we inject a fault in the longitudinal velocity of the ego vehicle, which is transferred to the ACC such that the ACC does not accurately know the vehicle's speed. This hazardous situation can lead to unintended acceleration as visualised and classified with the SAHARA methodology.

3   ML-Based FI Within the SAHARA Methodology

The Simulation-Aided Hazard Analysis and Risk Assessment (SAHARA) methodology focuses on assisting with (semi-) automated reasoning about the hazards and risks present in a safety-critical system [21]. In summary, this

---

[2] The model is an adapted version of https://www.mathworks.com/help/mpc/ug/adaptive-cruise-control-using-model-predictive-controller.html.

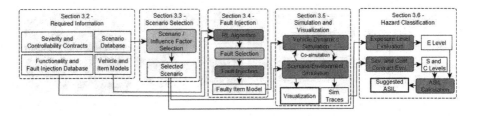

Fig. 3. The overall SAHARA architecture and workflow (adapted from [21]). Yellow blocks are automatic actions, and gray blocks are manual actions. (Color figure online)

methodology utilises specifications of scenarios, faults, and vehicle dynamics, which are combined and fed as input into simulations, which in turn provide data for visualisations and classification of the (potentially) hazardous situation.

The following sections will address applying the five prominent components of the approach as indicated in Fig. 3: *required information, scenario selection, fault injection and reinforcement learning algorithm, simulation and visualisation,* and *classification of hazard level.*

3.1 Required Information

Scenario Database. This scenario information combines the map and path information for vehicles, along with different *influence factors,* such as relevant characteristics of the vehicle, road conditions, and vehicle/pedestrian interactions [13,21].

Vehicle and Item Models. The SAHARA methodology requires detailed vehicle models, including the dynamics and the component(s) under study. Faults are injected into these component models, and the dynamics models are then used to simulate a scenario and assess the safety of the faulty component. In this work, we utilise the simple vehicle dynamics model built into CARLA, and the ACC model available within the Simulink documentation (see Sect. 2).

Functionality and Fault Database. Information on which component is under study and how it may fail is also necessary for utilising the SAHARA methodology, such that faults can be appropriately applied to the item models as discussed in Sect. 3.3. The functionality under study in this work is the reporting of the ego vehicle's longitudinal velocity to the ACC (see Sect. 2), with *sensor noise* and *stuck-to* faults available [21].

Severity and Controllability Contracts. The SAHARA methodology requires contracts to suggest levels of severity and controllability as defined by the ISO 26262 standard. These contracts are discussed in detail in Sect. 3.5.

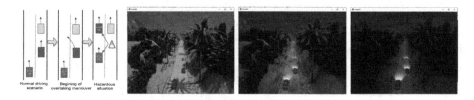

Fig. 4. Base scenario, and visualized as *DryRoad*, *RainyRoad*, and *NightRoad*. (Color figure online)

3.2   Scenario Selection

An analysis of automotive hazards involves reasoning about the safety of a component in various scenarios. These scenarios involve the layout of the roads, the road surface, as well as effects like the weather. These factors may all impact the dynamics or controllability of the vehicle and thus must be considered in a safety assessment process.

As envisaged in the SAHARA methodology [21], the safety engineer would select representative scenarios of interest through a tabular scenario description file. This work selects three scenarios: a straight road in clear weather, rainy weather, and night-time rainy weather, as shown in Fig. 4. The rainy weather affects both the visuals of the situation and the friction parameters of the road[3]. The night-time scenario does not affect the vehicle dynamics, but would make the situation more hazardous for a human driver.

Each scenario represents the same driving manoeuvre as shown on the left-hand side of Fig. 4. The red vehicle at the bottom is the ego vehicle, which must detect the blue middle vehicle moving into the left lane to overtake the furthest green vehicle. As explained in Sect. 2, the ACC must function appropriately on the ego vehicle to detect this movement into its lane, decrease the ego vehicle's speed if necessary, and avoid an accidental rear-end collision.

3.3   Fault Injection and Reinforcement Learning Algorithm

Fault injection (FI) is a well-known technique that exposes the system to a fault to allow the test engineer to understand if the system can adequately respond or whether further design changes are required. Our approach focuses on the most common *stuck-to-value* fault type in the sensors (input values) of components [25].

For example, a *stuck-to-value* fault may force a signal's value to be intermittently 'stuck' to a certain value at some simulation time, as modelled in Fig. 5. The middle block is a 'switch' block, which changes the output from the regular input I to the faulty value V at all timesteps after time T. However, this fault must still be parameterised to answer a) *where* to inject this fault, b) *when* the switch should occur, and c) *what* the faulty value should be [3].

---

[3] Friction values sourced from Fig. 24 of Singh and Taheri [28].

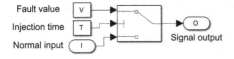

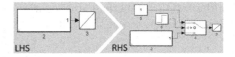

Fig. 5. *Stuck-to-value* fault at time T.   **Fig. 6.** Model transformation FI [19].

Fault Injection in the Adaptive Cruise Control. The ACC controls the ego vehicle's acceleration based on the information from incoming radar combined with the ACC's information on the longitudinal velocity of the ego vehicle (Sect. 2). As in [18], this work studies the presence of a stuck-to-value fault in the *ego longitudinal velocity* sensor of the ACC as seen in the top-left of Fig. 2. That is the sensor which reports the current velocity of the ego vehicle to the ACC. With this fault, the ACC will have incorrect knowledge of the ego vehicle's velocity, potentially resulting in a hazardous situation or collision.

In the SAHARA methodology, the safety engineer would select the longitudinal velocity signal as the fault location and the stuck-to-value fault from the ACC functionality and fault database. This FI is then performed using a framework to perform rule-based transformations on Simulink models [7,19]. Patterns utilising Simulink blocks can be matched in a model, and a rewrite pattern can then add, remove, or modify blocks. In Fig. 6 the left-hand side of the rule is the block pattern to match and the right-hand side is the replacement pattern.

Use of Machine Learning for Fault Injection. As a contribution to the SAHARA methodology, this work integrates the machine learning-based FI approach from previous work [18]. In this approach, a Reinforcement Learning (RL) framework searches the parameter space of the injected faults (*when* and *what value* to inject) over several simulations to force hazardous situations.

The framework utilises domain knowledge to set the boundaries and steer the direction of the parameter search for the RL agent. For example, the reward function in RL includes three parameters: the *time* of the simulation, the *velocity* of the ego vehicle, and the *relative distance* of the ego and lead vehicles. This reward function steers the search for fault values towards those that increase the velocity while decreasing the time until the collision occurs and the relative distance. This thus provokes as serious a crash as possible.

The FI framework runs multiple iterations of simulations with the fault parameters tuned each time to reach increasing reward function values corresponding to more hazardous situations. This results in a set of fault parameters to inject into the scenario to provoke the most hazardous behaviour found (as defined by reward value), and explore the most relevant safety consequences of faults.

Note that the RL framework must simulate the vehicle's behaviour in the scenario to determine the outcome. Currently, this simulation is performed 'headless' (without visualisation) within Simulink to avoid any overhead. These optimisation simulations are separate from the simulation required for visualisation as discussed in the next section.

3.4 Simulation and Visualisation

At this step in the SAHARA methodology, the appropriate scenario and fault (parameterised using RL) have been selected, and the fault has been injected into the component model. The next stage is the simulation of the scenario to produce a) a visualisation for use by experts in a safety assessment process, and b) traces for a preliminary safety classification.

As in [30], we select the open source CARLA simulator [8] for its high-quality visualisations, easy integration with other tools such as Simulink and Python, and default vehicle dynamics model. The scenarios are loaded into CARLA by modifying the weather and time of day on a built-in map.

The scenario is a co-simulation between CARLA and Simulink using a Python bridge to synchronise each time step. This simulation runs for a predetermined time as set in the scenario parameters. Simulink simulates the ACC, while CARLA simulates the vehicle dynamics and environment. Simulation traces are also produced for the assessment of hazardous situation. These signals include the *acceleration*, *velocity*, and a measure of the *collision impulse* of the ego vehicle (with any kind of object), the *relative velocity and distance* to the lead vehicle, and a measure of the *time gap* before a collision would occur [26].

From the simulation, CARLA also produces a visualisation for the safety engineer to utilise to reason about the consequences of the studied fault. These visualisations thus show to the safety engineer a sense of the driver's experience as well as possible outcomes of the failure of the component. This provides insight into how faulty behaviour could be hazardous.

3.5 Hazard Classification

The last step of the SAHARA methodology is to examine the resulting simulation traces to suggest a preliminary classification of the scenario hazard level.

The ISO 26262 standard specifies that an Automotive Safety Integrity Level (ASIL) be produced for a particular scenario by classifying the *exposure* (E), *severity* (S), and *controllability* (C) level of the hazardous scenario [11]. The ASIL is then provided by lookup in a table taking the E, C, and S into account to give ASIL QM (lowest) or A to D (highest). The ASIL, therefore, provides a guide to the safety-critical nature of a component's faults.

Exposure - The *exposure* (E) level of the scenario estimates the likelihood of the scenario from a scale from E1 (low exposure) to E4 (high exposure). The literature presents an automated approach to this calculation based on the probability of each influence factor in the scenario [13].

Severity - The *severity* (S) level concerns the potential injuries or death caused in the scenario, ranging from S1 (no injuries) to S4 (multiple severe injuries or deaths).

```
Contract sahara_sev_s2_a{
  description "S2: col_accel >= 40 G for 0.1"
  statements{
    Property col_accel :=
      div(collImpulse, constant CI_SCALE) >=
        val 40 gForce
  }
  scope Globally
  pattern Existence: col_accel occurs-
    eventually for-duration 0.1 s
  action Severity == S2
  generate-STL
}
```

```
Contract sahara_con_rt2{
  description "Time gap between vehicles is
    [0.5 1.3) seconds"
  statements{
    Property short_time_gap :=
      timeGap in Interval [ min 0.5 max 1.3
        unit second)
  }
  scope Globally
  pattern Existence: short_time_gap occurs-
    eventually
  action Controllability == C2
  generate-STL
}
```

Fig. 7. *S2 Severity* contract.   **Fig. 8.** *C2 Controllability* contract.

Table 1. *Severity* and *Controllability* contracts for hazard classification.

S. Level	Conditions	C. Level	Conditions
S0	$collImpulse = 0$ G	C0	$accel$ in $\pm 1.47\,\mathrm{m/s^2}$
S1	$collImpulse \geq 0.01$ G		$timeGap > 2.6$ s
S2	$collImpulse \geq 40$ G for 0.1 s	C1	$accel$ in $\pm 3.07\,\mathrm{m/s^2}$
	$collImpulse \geq 25$ G for 0.2 s		$timeGap$ in 1.3 to 2.6 s
	$collImpulse \geq 15$ G for 0.6 s	C2	$accel > \pm 3.07\,\mathrm{m/s^2}$
S3	$collImpulse \geq 100$ G for 0.01 s		$timeGap$ in 0.5 to 1.3 s
	$collImpulse \geq 50$ G for 0.04 s	C3	$timeGap < 0.5$ s
	$collImpulse \geq 45$ G for 0.1 s		
	$collImpulse \geq 30$ G for 0.3 s		
	$collImpulse \geq 25$ G for 0.8 s		

Controllability - The *controllability* (C) level of a situation also ranges from C1 to C4. This level represents the difficulty in controlling or avoiding the situation. This includes the driver of the vehicle as well as other participants in the scenario, such as pedestrians.

Controllability and Severity Contracts. The SAHARA methodology proposes assigning severity and controllability levels by developing temporal logic contracts that operate over the traces output from a simulation [21]. In this work, we further develop the provided contracts in Signal Temporal Logic (STL) [15] and apply them to the simulation traces produced by CARLA. Temporal logic is utilised due to the requirement to reason about both the value of signals as well as the temporal duration of conditions, such as 'at least 0.4 s'.

Severity Contracts - The ego vehicle collision impulse experienced in a collision can be used as a proxy for the severity level experienced [21]. Table 1 includes the assignment of collision impulse ranges to severity levels specified by [21] utilising data from [27]. For example, if the collision impulse is experienced at over 40 Gs of force for at least 0.1 s, then *S2* is assigned by Fig. 7.

Controllability Contracts - Two factors are included as proxies for controllability in Table 1: a) the longitudinal acceleration of the ego vehicle, where extreme values indicate a more aggressive driver [2], and b) a measure of reaction time for a driver in case of a lead vehicle emergency brake. That is, how many seconds the driver has to respond before a collision [26]. For example, Fig. 8 assigns *C2* (moderately uncontrollable) if the time gap between the vehicles is less than a standard reaction time of 1.3 s [5].

Contract Evaluation. For verification, each contract in Table 1 is mapped to an equivalent STL representation [4]. This mapping process increases the usability of the contract verification approach, as STL can be difficult to reason about and write by hand. Instead, this domain-specific contract language allows for the specification of contracts using familiar operators and units.

For example, Eq. (1) shows the STL generated for the *Severity* contract seen in Fig. 7. This STL converts the collision impulse into the correct unit and implements the *Existence* pattern with duration present in Fig. 7. The resulting STL is then verified against the vehicle dynamics traces produced by CARLA.

$$\textbf{eventually}(\textbf{always}[0:0.1]((collImpulse/14.0) \geqq 392.0)) \tag{1}$$

These traces and the contract STL are fed as input into the Python-based RTAMT verification library[4]. Each contract's STL is checked in turn in an offline manner on the simulation traces. If the specification succeeds, then the trace (and therefore the situation) is assigned at least that severity or controllability level.

Outcome. As discussed in the SAHARA methodology and Sect. 3.2, the exposure (E) level can be determined by examining the influence factors of the scenario. The severity (S) and controllability (C) levels are then calculated by verifying temporal logic contracts on the resulting simulation trace. The resulting ASIL is then determined through a lookup table of the E, S, and C levels.

This ASIL suggestion is presented to the safety engineer along with a visualisation of the fault scenario to assist in assessing the hazardous situation. Even though these artifacts are only a suggestion of possible outcomes, they may provide insight into the hazardous nature of the situation.

4   Results

Table 2 displays the results of our application of the SAHARA methodology with ML-based FI to the ACC use case. For each scenario, the calculated *Severity* and *Controllability* levels and a suggested ASIL are presented. As each of these scenarios is quite likely, an exposure level of *E4* is statically assigned. Visualisations are also available online for all regular and faulty scenarios[5].

For the non-faulty scenarios, the simulation traces indicate that no collision occurred. Therefore the severity level is the lowest, and the ASIL remains as

---

[4] https://github.com/nickovic/rtamt

[5] https://www.youtube.com/playlist?list=PLNyNvnuIvPKvsmUT1I-hwEYDMyZ7 YGZkA.

Quality Management (QM). In the *RainyRoad* scenario, the stopping distance and the *timeGap* between the vehicles are modified due to the lowered road friction. The contracts from Table 1 thus assign a higher *Controllability* level (less controlled).

Table 2. Suggested hazard classifications and visualisations for all scenarios.

Scenario	Without fault			With fault		
	DryRoad	*RainyRoad*	*NightRoad*	*DryRoad*	*RainyRoad*	*NightRoad*
S Level	0	0	0	3	3	3
C Level	1	2	2	3	3	3
ASIL	QM	QM	QM	D	D	D

For the scenarios with faults, a collision is provoked by the machine learning approach in Sect. 3.3. There is a significant collision impulse as the velocity sensor of the ego vehicle becomes faulty as the lead vehicle is overtaking. The rapid acceleration then causes an impact between the vehicles that could cause grave injury. Thus, a severity level of *S3* is assigned, leading to a suggested hazard classification of ASIL D (the highest level).

The simulation time in our approach is divided into three parts: a) injecting faults into the ACC Simulink model takes around 2.7 min, b) The RL process then operates on this faulty model and finds multiple parameter sets (fault amplitude and injection time) causing hazardous situations in about 13.5 min, and c) For each one of these critical parameter sets found, simulation and visualisation takes around 2.5 min with three seconds for assigning an ASIL. Therefore, an end-to-end run of this methodology takes about 20 min.

The FI process and the CARLA simulation took place on a 32-core processor running at 2.99 GHz with 24 GB of memory and a graphics card with 11 GB of memory. Only the RL part is multi-threaded. The contract verification process took place on a 12-core Intel i7-8850H CPU at 4.3 GHz with 16 GB of memory.

5   Discussion

Approach Benefits. The approach of the SAHARA methodology (Sect. 3) is to ease the effort required by the safety engineer in performing a safety assessment of the component. This process can (semi-)automatically produce simulations and visualisations of component faults.

Thus our addition of ML-based FI is a natural step, as it attempts to optimise the parameters of injected faults such that a more hazardous situation develops (Sect. 3.3). As in previous work [17], this approach is superior to random-based FI in exploring the fault space in terms of fault coverage and number of simulation to find the first critical fault. The addition of this step may thus allow the safety engineer to discover previously unknown situations where unsafe behaviour occurs and increase fault coverage. Therefore, the simulations and visualisations increase the safety engineer's comprehension of the possible component faults and offer concrete discussion points and insights.

A substantial value of this automatic SAHARA methodology with ML-based FI is the (semi-)automation, leading to new visualisations and classification results produced in a matter of minutes (Sect. 4), although a computationally powerful machine is required. We envision an assessment workflow where visualisations could be interactively produced during a safety assessment discussion about various scenarios or faults.

Human-in-the-Loop Necessary. At first glance, not having a fully automated framework and keeping a human-in-the-loop is a limitation for SAHARA. However, this is not possible in safety assessment, requiring a tremendous amount of experience and in-depth domain knowledge that is unlikely to be adequately captured by an automaton. The safety engineer's role cannot be replaced entirely, despite the cost of safety assessment discussions. This is due to barriers, including legal responsibilities or insufficient simulation fidelity.

Approach Limitations. Our approach inherits the weakness of ML. For example, the time taken to search for a hazardous situation depends on the proper modelling of the vehicle, environment, and the reward function to steer the search [10,18]. If the representation of the problem or the reward function is insufficient, then a hazardous situation may not be found. The probabilistic nature of ML also means that it cannot be predicted at what time a hazardous situation will be found. Another issue is that this searching process is computationally expensive and can take a significant amount even on a powerful computer. As the vehicle and component models become more realistic and even more complex, this could limit the applicability of this technique.

Here we suggest two challenges not addressed in this work:

a) For any particular scenario, the situation found by the SAHARA methodology may not be the absolute worst-case due to the infinite and granular space of situations possible. For example, a steering fault may or not cause a head-on crash based on a margin of centimetres or less. Instead, this SAHARA methodology aims to present a representative simulation, such that a safety engineer can recognise the inherent danger in this situation.
b) An extended assessment of controllability and severity levels involves more than just vehicle dynamics models [13,21]. For example, controllability relies on the driver's reaction (and others), which requires modelling of driver's behaviour [23]. Likewise, severity depends on the dynamics of the driver inside the vehicle, such as possible whiplash injuries or interactions with airbags.

Due to these challenges, we restrict our application of the SAHARA methodology to only providing *representative* hazardous scenario visualisations and hazard classification *suggestions* to assist safety assessments for a safety engineer.

6 Related Work

Assessing safety in the automotive domain is an active field of research, especially in assessing autonomous vehicles [22]. For example, the open-source MOBATSim framework combines a sub-microscopic vehicle simulator with reasoning about faults [24]. Faults are injected into the front distance sensor for vehicles in a platooning scenario, with a Monte Carlo approach to find more hazardous situations. Results are presented by the framework to relate the fault parameters with an indication of safety specifications violations. This approach also addressed the design of safety-critical systems, where faults and scenarios are evaluated for two design variants to choose the safer design [26].

Juez *et al.* examine the role of FI in the context of the ISO 26262 standard and its safety assessment process [12]. The SABOTAGE framework is defined, which performs FI on a lateral control vehicle component, simulates the faulty scenario versus a non-faulty (golden) scenario, and then determines the maximum time that a fault can be present in the system.

In this work, we utilise RL to adjust the fault parameters to provoke a more serious situation [18]. Alternatively, the work of Althoff and Lutz [1] adjusts the scenario parameters themselves, such as doubling the initial speed of the ego vehicle in a scenario or arranging the movement of vehicles to block off lanes or increase the danger of overtaking.

Duracz *et al.* explore the use of *rigorous simulations* to assign severity levels in an ISO 26262 safety assessment context [9]. Rigorous simulations operate on explicit dynamics models and produce provably correct bounds for the behaviour. As in our work, Duracz *et al.* base the severity level on the change in the velocity signal upon collision.

Tuncali *et al.* define STL specifications for both system- and component-level to be proved on a simulation [29]. An example is that when an object is *visible* to sensors, the object must be *detected* by the sensors within a specific time frame. An optimisation framework is then employed to find scenarios that falsify the specifications. In contrast, our work performs the optimisation on the FI to search towards the most hazardous situation, and the specifications are only for hazard classification.

Zapridou *et al.* mirror our work by presenting STL verification of properties on an ACC use case using the CARLA simulator [30]. However, our work focuses on the FI portion of determining safety, and also places the intelligent FI within the SAHARA safety assessment process.

7 Conclusion and Future Work

This work has presented the addition of machine learning-based Fault Injection (FI) to the Simulation-Aided Hazard Analysis and Risk Assessment (SAHARA) methodology, as demonstrated on a safety-critical use case of an Adaptive Cruise Control (ACC). Specifically, Reinforcement Learning (RL) explores the parameters of faults injected into the ACC such that a hazardous situation is provoked. This situation is then simulated in the open-source automotive simulator

CARLA [8] to produce a visualisation as well as simulation traces for use in indicating the hazard classification level of the situation. Example situations are shown to demonstrate the applicability of our approach, and timing results indicate that this approach is relatively interactive as it takes only around twenty minutes to complete end-to-end, and less than three minutes to produce a new visualisation.

The natural extension of this work is to validate it within an industrial safety assessment process. In particular, performing a study following the safety engineers as they perform the standard hazard analysis and risk assessment procedure, and then comparing this with our proposed SAHARA with RL-based FI. Metrics and user surveys would then indicate the time saved and satisfaction with the (semi-)automated approach.

References

1. Althoff, M., Lutz, S.: Automatic generation of safety-critical test scenarios for collision avoidance of road vehicles. In: 2018 IEEE Intelligent Vehicles Symposium (IV), pp. 1326–1333. IEEE (2018)
2. Bae, I., Moon, J., Seo, J.: Toward a comfortable driving experience for a self-driving shuttle bus. Electronics 8(9), 943 (2019)
3. Benso, A., Prinetto, P.: Fault Injection Techniques and Tools for Embedded Systems Reliability Evaluation, vol. 23. Springer, Boston (2003). https://doi.org/10.1007/b105828
4. Bernaerts, M., Oakes, B., Vanherpen, K., Aelvoet, B., Vangheluwe, H., Denil, J.: Validating industrial requirements with a contract-based approach. In: 2019 ACM/IEEE 22nd International Conference on Model Driven Engineering Languages and Systems Companion (MODELS-C), pp. 18–27. IEEE (2019)
5. Coley, G., Wesley, A., Reed, N., Parry, I.: Driver reaction times to familiar, but unexpected events. TRL published project report (2009)
6. Coppola, R., Morisio, M.: Connected car: technologies, issues, future trends. ACM Comput. Surv. 49(3), 1–36 (2016)
7. Denil, J., Mosterman, P.J., Vangheluwe, H.: Rule-based model transformation for, and in Simulink. In: Proceedings of the Symposium on Theory of Modeling & Simulation-DEVS Integrative, pp. 1–8 (2014)
8. Dosovitskiy, A., Ros, G., Codevilla, F., Lopez, A., Koltun, V.: CARLA: an open urban driving simulator. In: Proceedings of the 1st Annual Conference on Robot Learning, pp. 1–16 (2017)
9. Duracz, A., et al.: Advanced hazard analysis and risk assessment in the ISO 26262 functional safety standard using rigorous simulation. In: Chamberlain, R., Edin Grimheden, M., Taha, W. (eds.) CyPhy/WESE -2019. LNCS, vol. 11971, pp. 108–126. Springer, Cham (2020). https://doi.org/10.1007/978-3-030-41131-2_6
10. Hauer, F., Pretschner, A., Holzmüller, B.: Fitness functions for testing automated and autonomous driving systems. In: Romanovsky, A., Troubitsyna, E., Bitsch, F. (eds.) SAFECOMP 2019. LNCS, vol. 11698, pp. 69–84. Springer, Cham (2019). https://doi.org/10.1007/978-3-030-26601-1_5
11. International Organization for Standardization: ISO 26262: Road vehicles-functional safety (2011)

12. Juez, G., Amparan, E., Lattarulo, R., Rastelli, J.P., Ruiz, A., Espinoza, H.: Safety assessment of automated vehicle functions by simulation-based fault injection. In: 2017 IEEE International Conference on Vehicular Electronics and Safety (ICVES), pp. 214–219. IEEE (2017)
13. Kemmann, S.: SAHARA-a structured approach for hazard analysis and risk assessments. Ph.D. thesis, Fraunhofer-Institut für Experimentelles Software Engineering (2015)
14. Macher, G., Sporer, H., Berlach, R., Armengaud, E., Kreiner, C.: SAHARA: a security-aware hazard and risk analysis method. In: 2015 Design, Automation & Test in Europe Conference & Exhibition (DATE), pp. 621–624. IEEE (2015)
15. Maler, O., Nickovic, D.: Monitoring temporal properties of continuous signals. In: Lakhnech, Y., Yovine, S. (eds.) FORMATS/FTRTFT -2004. LNCS, vol. 3253, pp. 152–166. Springer, Heidelberg (2004). https://doi.org/10.1007/978-3-540-30206-3_12
16. Meyers, B., Gadeyne, K., Oakes, B.J., Bernaerts, M., Vangheluwe, H., Denil, J.: A model-driven engineering framework to support the functional safety process. In: 2019 ACM/IEEE 22nd International Conference on Model Driven Engineering Languages and Systems Companion (MODELS-C), pp. 619–623, September 2019
17. Moradi, M., Oakes, B., Denil, J.: Machine learning-assisted fault injection. In: 39th International Conference on Computer Safety, Reliability and Security (SAFECOMP), Position Paper, Lisbon, Portugal (2020)
18. Moradi, M., Oakes, B.J., Saraoglu, M., Morozov, A., Janschek, K., Denil, J.: Exploring fault parameter space using reinforcement learning-based fault injection. In: 2020 50th Annual IEEE/IFIP International Conference on Dependable Systems and Networks Workshops (DSN-W), pp. 102–109. IEEE (2020)
19. Moradi, M., Van Acker, B., Vanherpen, K., Denil, J.: Model-implemented hybrid fault injection for Simulink (tool demonstrations). In: Chamberlain, R., Taha, W., Törngren, M. (eds.) CyPhy/WESE -2018. LNCS, vol. 11615, pp. 71–90. Springer, Cham (2019). https://doi.org/10.1007/978-3-030-23703-5_4
20. Polydoros, A.S., Nalpantidis, L.: Survey of model-based reinforcement learning: applications on robotics. J. Intell. Robot. Syst. **86**(2), 153–173 (2017). https://doi.org/10.1007/s10846-017-0468-y
21. Rafael, A.B.J., Bachir, Z.: SAHARA: Simulation aided hazard analysis and risk assessment methodology. In: Risk Analysis XII, vol. 129, p. 41 (2020)
22. Riedmaier, S., Ponn, T., Ludwig, D., Schick, B., Diermeyer, F.: Survey on scenario-based safety assessment of automated vehicles. IEEE Access **8**, 87456–87477 (2020)
23. Salvucci, D.D.: Modeling driver behavior in a cognitive architecture. Hum. Factors **48**(2), 362–380 (2006)
24. Saraoglu, M., Morozov, A., Janschek, K.: MOBATSim: Model-based autonomous traffic simulation framework for fault-error-failure chain analysis. IFAC-PapersOnLine **52**(8), 239–244 (2019)
25. Saraoğlu, M., Morozov, A., Söylemez, M.T., Janschek, K.: ErrorSim: a tool for error propagation analysis of Simulink models. In: Tonetta, S., Schoitsch, E., Bitsch, F. (eds.) SAFECOMP 2017. LNCS, vol. 10488, pp. 245–254. Springer, Cham (2017). https://doi.org/10.1007/978-3-319-66266-4_16
26. Saraoğlu, M., Shi, Q., Morozov, A., Janschek, K.: Virtual validation of autonomous vehicle safety through simulation-based testing. In: 20. Internationales Stuttgarter Symposium. P, pp. 419–434. Springer, Wiesbaden (2020). https://doi.org/10.1007/978-3-658-29943-9_33
27. Shanahan, D.F.: Human tolerance and crash survivability. In: Pathological Aspects and Associate Biodynamics in Aircraft Accident Investigation (2004)

28. Singh, K.B., Taheri, S.: Estimation of tire-road friction coefficient and its application in chassis control systems. Syst. Sci. Control Eng. **3**(1), 39–61 (2015)

29. Tuncali, C.E., Fainekos, G., Prokhorov, D., Ito, H., Kapinski, J.: Requirements-driven test generation for autonomous vehicles with machine learning components. IEEE Trans. Intell. Veh. **5**(2), 265–280 (2019)

30. Zapridou, E., Bartocci, E., Katsaros, P.: Runtime verification of autonomous driving systems in CARLA. In: Deshmukh, J., Ničković, D. (eds.) RV 2020. LNCS, vol. 12399, pp. 172–183. Springer, Cham (2020). https://doi.org/10.1007/978-3-030-60508-7_9

Safety Validation and Simulation

SASSI: Safety Analysis Using Simulation-Based Situation Coverage for Cobot Systems

Benjamin Lesage and Rob Alexander[(✉)]

Department of Computer Science, University of York, York, UK
{benjamin.lesage,rob.alexander}@york.ac.uk

Abstract. Assessing the safety of collaborative robot (cobot) systems is a difficult task due to the myriad of possible interactions between robots and operators, and the potential for injury to the operators. Using a situation coverage approach we can define the individual components of such interactions, and thereby describe the problem space and the coverage achieved when testing it. In this paper, we propose a situation coverage approach for testing the safety of a cobot system. Our approach suggests using a combination of safety analysis techniques and simulation-based testing to define situations of interest and explore hazardous situations while only endangering virtual operators. We challenge our assumptions by applying our method to an example based on a real-world use-case. The proposed metrics, if they provide no advantage to guided test generation techniques over random ones, helped us trim the generated configuration landscape to identify safety gaps.

Keywords: Cobot · Situation coverage · Simulation-based testing · Safety analysis

1 Introduction

Cooperative robots (cobots) [4] aim to allow human operators and robot workers to share the same work-space, and work jointly to achieve a common goal. Because of safety concerns, however, users tend to build physical barriers to isolate robots from the operators [19,26]. These barriers tend to limit the level of cooperation between human and robot, and reduce the advantages of deploying cobots in the industrial space.

The CSI: Cobot project [7] aims to reduce the need for barriers in cobot systems. The project proposes novel sensing and control techniques to improve cobots' awareness of their environment, especially regarding interactions with human operators. A crucial requirement for the adoption of new techniques is achieving confidence in the overall safety of the system. We consider safety aspects in the context of the CSI: Cobot project, in particular investigating the impact of changes in the system setup.

© Springer Nature Switzerland AG 2021
I. Habli et al. (Eds.): SAFECOMP 2021, LNCS 12852, pp. 195–209, 2021.
https://doi.org/10.1007/978-3-030-83903-1_13

Simulation-based techniques, e.g. the CARLA simulator [8], are a common approach to evaluate autonomous systems [5,21,29]. Simulations allow fast iterations over varied configurations, including hazardous ones, without endangering the system itself, or its environment. However, the use of simulations for testing safety constraints raises important issues. First, the safety case must ensure the simulated environment is representative of the system under consideration [2,16]. Second, the tools must decide on a set of configurations (from the effective-infinite configuration space) to drive testing and evaluate confidence in the system safety [18].

Our approach relies on simulation-based, situation-driven testing to evaluate the safety of a cobot system. It builds on the safety analysis of the cobot system under consideration, which identifies accidents and losses that arise from unsafe operation. The safety artefacts, generated as part of this analysis, capture undesirable situations which though not hazardous by themselves may lead to a loss. In the context of situation coverage [3], safety situations inform our exploration of the system configurations and provide for an evaluation of the confidence in the system safety.

We first introduce a manufacturing cobot use case to highlight our approach in Sect. 2. We then outline the general principles of our approach in Sect. 3. Section 4 discusses the identification of safety analysis artefacts, then monitored to identify hazards and guide the generation of simulation configuration, respectively in Sect. 5 and 6. Section 7 introduces our setup to assess the validity of our approach. We compare our approach to existing work (Sect. 8) before summarising our results in Sect. 9.

2   Case Study: Industrial Manufacturing Cobot

We consider an industrial use case, defined in the context of the CSI: Cobot project [7], involving the cooperation of a human operator and a robotic arm to assemble small metal components ("assemblies"). Note that the general principles of our approach are not tied to the specific use case or tools we discuss in the following. The operator provides a non welded assembly at a designated work bench. The arm then retrieves the assembly and transports it to a spot welder for processing, before returning to the same work bench for a handover. The operator should keep out of the cell while the arm is active less he puts himself at risk of injury. All the processing currently occurs within a walled cage, with sensors to ensure no operator is present while the welder is active or the arm is moving. The cell is depicted in Fig. 1a.

Our default setup is an abstraction of the industrial use-case, depicted in Fig. 1b. All major components are in place, the welder, the cobot arm, and the shared bench. The highlighted space in the middle defines the cell region. The safety cage has been omitted. In lieu of a cage, a presence sensor stops the arm when the operator enters the cell. Due to constraints of our simulation environment, control of the arm and operator is limited with no exchange of assembly between them. The arm is programmed to loop between two waypoints from the bench at the bottom, to the welder at the top.

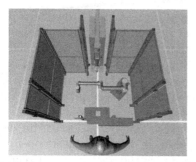 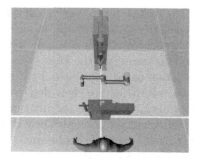

(a) Cell configuration with an operator (bottom), and an arm (centre) with a clear path to a welder (top).

(b) Setup under evaluation, with the cell region outlined (center) and the safety walls removed.

Fig. 1. Considered industrial use case configuration and evaluated setup.

3   Overview of the SASSI Method

The SASSI method relies on the artefacts produced by the safety analysis. A key principle of our approach is to derive sufficient knowledge from the safety artefacts to guide the testing of the system. The objective is to understand if and how safety issues might arise in the system. Safety artefacts provide a safety-centric view of the entities involved in the system, their interactions, their relevant properties, and how those may lead to hazards; the safety analysis thus informs multiple aspects of the toolchain.

The system design is at the root of the process and it guides all further steps, defining the environment, its operating conditions, and safety requirements. Our method is building confidence into the system by testing it, searching for safety-relevant configurations, to provide feedback regarding safety aspects into the design. Figure 2 presents the overall workflow of the analysis method.

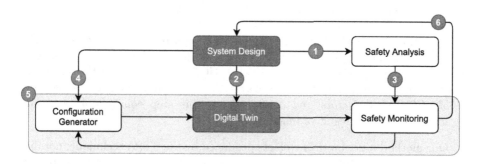

Fig. 2. Overview of the analysis method

The first step is to perform a safety analysis of the system (1) to understand the occurrences of hazards in the system, and the conditions leading to such events (Sect. 4). The system design further informs the development of a simulation environment, the Digital Twin (2) which constitutes the baseline for our simulation-based approach. The Digital Twin also allows for the evaluation of system setups before their deployment in the actual cell. The artefacts of the safety analysis provide information on safety-relevant situations in the system (3) and the conditions for their detection (Sect. 5). The system design further constrains acceptable configurations of the system. It defines elements open to variations and their degree of freedom, thus outlining the domain of our search (4). Safety artefact components observed running the simulation, from a generated configuration, provide feedback on situations of interest (5) during the search (Sect. 6). Finally, the coverage of generated and observed situations during analysis (6) provides some confidence in the safety of the system, or highlights shortcomings that need to be addressed.

4   Analysing the System Safety

The Safety Analysis aims to understand the safety of the system, by identifying potential hazards, and the situations or causes leading to these events so they can be managed. The analysis needs to be aware of the components and entities interacting in the system as well as its operating environment. Without loss of generality, we present the Systems Theoretic Process Analysis (STPA) technique as the underlying safety analysis. The results of the application of STPA to our use case and our experience have been documented in [1].

STPA [20] originates from systems approaches to safety engineering. Accidents are assumed to arise from insufficient feedback or inadequate control in the system, as modelled by a control structure. The STPA analysis process is as follows:

1. *Identify accident and loss scenarios*: these encompass a range of undesired events such as damage to property, injury to humans, or environmental pollution.
2. *Construct the system control structure*: the control structure captures the entities in the system and the flow of control and feedback between them.
3. *Identify unsafe control actions (UCAs)*: UCAs correspond to the execution of actions in undesirable configurations of the environment and the system.
4. *Identify causal factors and control flaws*: this step considers how unsafe actions arise as a result of inadequate control.

Our method relies on the artefacts produced by the STPA technique, notably hazards and unsafe control actions. Hazards identify events which by definition challenge the safety of the system. Unsafe control actions (UCAs), while not hazardous themselves, are undesirable. A UCA is an action the execution of which or lack thereof, in a given a configuration of the system, may give rise to a hazard. Each safety artefact thus captures a situation, a combination of components relevant to the safety of the system.

Our intuition is that monitoring for UCAs during testing can help identify hot spots for safety in the explored space of configurations, and focus the analysis effort on those regions more likely to result in hazards. The identification and monitoring of individual situation components can further guide the search towards UCA occurrences. Situation components finally outline a coverage target for the search. Focus should be given to strategies which cover varied combinations of situations, as they provide more confidence in the safety of the system.

Example 1 *We consider the cell in Fig. 1a and two situations captured by the safety analysis. Should the arm move while its path is obstructed (UCA **MUCA-1**), it may lead to a situation where a hazardous collision occurs involving the arm (Hazard **mH-3**). Configurations where there is an **Obstruction in cell** and the **Arm is moving** are more likely to trigger these safety-relevant situations.*

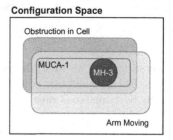

5   Monitoring Safety Artefacts

Monitoring aims to identify the set of safety-relevant situations that occurred during simulation. The simulation environment tracks the state of relevant components over time, producing a trace of events in the system. Runtime Verification methods [10] provide a vast array of tools and techniques to identify the violation of specific properties in a system from such a trace.

We use Fuzzy Linear Temporal Logical (LTL) [9,12] to model safety artefacts. Our intuition is to measure how close an artefact is to occurring over time in a given configuration. An LTL formula captures a condition on the future of a path, combining predicates using logic, e.g. a or b, and temporal operators, e.g. d *eventually* occurs. The truth value of a predicate under fuzzy valuation [12] ranges between 0 (false) and 1 (true). As opposed to a crisp *true* or *false* valuation, fuzziness should provide search heuristics with a finer-grain metric to compare different configurations. We further extend value comparison operators with a tolerance. Once the compared values are in the tolerance range, the comparison valuation linearly tends towards 1 as they draw closer to each other.

Example 2 *Hazard MH-2 captures the situation where the arm exceeds its velocity restrictions in proximity of another entity. It is divided in two components.*

1. *the arm exceeds its velocity restriction:* $comp_s(t) = arm.velocity(t) \geq_V PROX_V$
2. *the arm is in proximity of an entity:* $comp_d(t) = arm.distance(t) \leq_D PROX_D$

arm.velocity(t) and arm.distance(t) respectively capture the velocity of the arm and its distance to the closest entity at instant t. Constant $PROX_V$ constrains the arm velocity in proximity of another entity, and $PROX_D$ defines the proximity distance threshold for the arm.

If the arm is in close proximity of an entity, $arm.distance(t) \leq PROX_D$, then the constraint is satisfied: $comp_d(t) = 1$. Conversely as per our fuzzy valuation rules, if the arm is far enough from the closest object, $\mathcal{D} + PROX_D < arm.distance(t)$, then constraint is not satisfied: $comp_d(t) = 0$. The valuation of $comp_d$ linearly increases as the difference between arm.distance and $PROX_D$ decreases.

*MH-2 is satisfied for a given configuration if both conditions, $comp_s$ and $comp_d$, **eventually** occur at the same time. Under fuzzy logic, the **conjunction** and **always** temporal operator (respectively **disjunction** and **eventually**) are defined as the maximum (respectively minimum) of the combined predicates. The valuation of MH-2 under configuration c can thus be computed as:*

$$Occurs(MH\text{-}2, c) = \max_t \Big(min\big(comp_d(t), comp_s(t)\big) \Big)$$

The situations we monitor for are formalised from STPA artefacts, namely Hazards, UCAs, and their components. The textual description of each artefact needs to be translated into a formal monitor specification, as per Example 2, which might introduce errors in the process. We rely on a validation step to assess the suitability of our formal artefact specification. Each formalised artefact is tested against a number of crafted event traces, e.g. setting the value of *arm.velocity* and *arm.distance* at different instants t, and compared to expected truth values.

The simulation needs to expose the required information for monitoring. The safety analysis thus informs the design of the simulation not only in terms of the components and actions that should be modelled, but also regarding some of the events that need to be tracked. The safety monitors can be incorporated in the simulation tool itself, or applied to the output of the tool. We rely on a separate, offline monitoring approach. This allows the parallel development of the simulation environment, and the definition and validation of monitors. Artefacts are formalised into expressions in the Python language using the MTL library [25] for monitoring.

The safety monitors might rely on ground truth information that is unavailable or incorrect in the physical system due respectively to sensing gaps or faults. Conversely, some artefacts may need to be simplified for evaluation and monitoring, preferably subsuming the target condition to prevent false negatives at the cost of false positives.

Example 3 *In the absence of a physics model, a collision between the arm, the operator, or the assembly can be registered as soon as they connect, irrespective of their relative speed and mass.*

6   Generating and Evaluating Configurations

Automated testing techniques for autonomous robots need to address a number of challenges. The tools need to evaluate the confidence in the safety of the system to assess whether or not sufficient testing has been performed. They should also guide the testing strategy to produce situations that are interesting but also plausible and realistic.

Situation coverage [3] is a coverage criterion adapted to the testing of autonomous robots. The underlying principle is to identify components of the environment the system might encounter, identify how they can vary, and ensure that they, and combinations thereof, are evaluated during testing. High coverage speaks for the quality of the testing strategy and provides an indicator that the system has been considered in a variety of contexts. Testing of an autonomous vehicle would be required to navigate different types of road intersection with various shapes, combined with the type of vehicles it might encounter at the intersection, their direction of travel, and any other reasonable factor.

Macro-level components define a requirement on the inputs and their combinations exercised during testing. Their identification is obviously informed by the system design which outlines reasonable operating configurations for the system. Micro situation components define the set of situations that should always, or never, be observed during the test campaign. The safety analysis provides such information, capturing the set of events the system should cope with, e.g. a human enters the cell during a welding operation.

We propose using the safety artefacts to guide testing. The occurrence of a safety artefact reflects negatively on the confidence in the safety of the system. We define a fitness metric to use as an objective function with state-of-the-art search heuristics. A configuration with a higher fitness triggers the occurrences of more artefacts and will be favoured by the search:

$$Fitness(c) = \sum_{h \in Hazards} W_H \times Occurs(h, c) + \sum_{u \in UCAs} W_U \times Occurs(u, c) \quad (1)$$

where $Hazards$ and $UCAs$ are the set of hazards and UCAs identified by the safety analysis, $Occurs(s, c)$ is the fuzzy truth value capturing the occurrence of artefact s under configuration c (see Sect. 5), W_H and W_U weigh the occurrences of hazards and UCAs. We consider $W_H = 10$ and $W_U = 1$ in the following although weights can be adjusted per artefact, based on their severity.

To assess the coverage achieved by test campaign t, we consider the portion of safety artefacts triggered by the configurations encountered during testing:

$$ArtefactCoverage(t) = \frac{|\{s | s \in Artefacts \wedge \exists c \in t, Occurs(s, c) = 1\}|}{|Artefacts|} \quad (2)$$

Where $Artefacts = Hazards \cup UCAs$ is the set of safety artefacts identified by the safety analysis.

As the absence of observed safety artefacts is not a guarantee of safety, we further split artefacts into their individual components to compute coverage metrics. The components coverage is a measurement of the completeness of the testing with regards to safety components. A component or a combination thereof must be observed in all possible states to be considered as fully covered:

$$ComponentCoverage_N(t) = \frac{\left| \bigcup_{s \in \binom{N}{SC}} \bigcup_{c \in t} Occurs^N(s,c) \right|}{2^N \times \left| \binom{N}{SC} \right|} \qquad (3)$$

Where $Occurs^N(s,c)$ captures the joint occurrences of components in s in configuration c. As an example consider components A and B. $Occurs^2(\{A,B\},c)$ can contains at most 4 values, $A \wedge B$, $A \wedge \neg B$, $\neg A \wedge \neg B$, $\neg A \wedge B$. SC is the set of safety components derived from $Artefacts$, and $\binom{N}{SC}$ the set of N-length combinations of safety components. SC is built from the safety artefacts by automatically collecting the predicates and comparisons in their formulation, e.g. $comp_d$ and $comp_s$ in Example 2.

7   Evaluation

The SASSI method outlines general principles for testing a cobot system for safety. We rely in particular on safety artefacts to capture safety-relevant situation components. Those components guide the search through the configuration space for interesting scenarios, and provide an evaluation of the relevant coverage of the situation space. This section challenges our intuitions:

– Research Question 1. Are UCA necessary for hazards to occur?
– Research Question 2. Are situation-based heuristics a good guide for testing?

7.1   Problem Space

We identified in Table 1a the artefacts monitored by the simulation. We simplified some conditions w.r.t. to the original artefacts to cope with limitations of the simulation environment, as suggested in Sect. 3. The configuration variables in Table 1b outline the configuration domain and search space.

7.2   Simulation Setup

We discuss the integration for our evaluation with the Digital Twin developed in the CSI: Cobot project [7]. The simulation captures a model of the system and its behaviour. It is considered as a black box in our approach, used to evaluate the system's response to varied situations without any risk to its actors. The simulation needs to satisfy the safety monitoring and configuration requirements.

The simulation scene is setup as described in Sect. 2. The position of the operator is fixed during each run, and the safety stop will be prevent the arm from moving if the operator is inside the cell. Collisions are recorded as soon as the arm and operator connects. Virtual sensors, with no pendent in the physical system, provide for ground truth information regarding the velocity of the arm and its proximity to other entities, namely the operator. The cell region is defined as a volume which monitors entities entering or leaving.

Table 1. Configuration of the simulation for evaluation

(a) Safety artefacts monitored by the simulation.

Id	Event
MH-1	The arm exceeds its velocity restriction in either region.
MH-2	The arm exceeds its velocity restriction in proximity of another entity.
MH-3	The arm, assembly, or operator is compromised because of a collision.
MUCA-1	The arm moves while the cell is obstructed.
MUCA-2	The arm stops moving before it reaches its target position.

(b) Definition of the problem search space

Variable	Type	Semantic
Arm position	$(i, j, k) \in \mathbb{R}^3$	Where the arm is in the cell, outlined in Figure 1b
Operator position	$(x, y, z) \in \mathbb{R}^3$	Where the operator is in the cell or outside
Restrict Velocity	$(b \in \mathbb{B})$	Likely velocity threshold violations if unset

The processing framework for our analysis focuses on interactions with the Digital Twin (illustrated by Step 5 in Fig. 2), ensuring valid configuration files are generated and observations can be processed to identify situations of interest. The Digital Twin provides hooks to configure pre-existing entities in the simulated environment. All entity properties can be controlled through a unified configuration file. Communication between independent actors within the twin occurs through message passing, similarly to the real platform. Listeners capture all cross-entity messages, functional or safety-related, issued during a run into in a unified record.

Example 4 *The cell region is defined as a safety region volume which monitors entities entering or leaving the volume. The arm safety stop discussed in Sect. 2 reacts to unexpected entrances upon receiving the corresponding message. The same message provides for the computation of the "obstruction in cell" component during safety monitoring.*

Our toolchain supports the generation of configuration files for the Digital Twin, exposes primitives to process the messages recorded during a run, and allows for the evaluation of properties on the resulting observations. We configure the process by defining the domain for the search, the set of safety artefacts under evaluation, and the conversion from messages to variables required for artefact evaluation such as *arm.distance* in Example 2. The tools automatically extract the required safety components, and the corresponding coverage and fitness metrics for a run. Combined with a search heuristic, they provide the automated evaluation of the system.

7.3 Search Heuristics

We consider a number of heuristics to explore the configuration space. All heuristics were provided with the same budget of 1000 runs and the same input domain.

Ran randomly explores the configuration space, generating solutions within the provided constraints. It is used as a baseline for comparison against guided approaches.

GA-max (resp. **GA-min**) uses an elitist genetic algorithm [24,27] to maximise (resp. minimise) the fitness metric introduced in Sect. 6. Genetic algorithms operate on a population of configurations by selecting, and mutating the best individuals. The two configurations respectively aim to maximise and minimise the *concurrent* occurrences of safety artefacts in the system.

Quality diversity algorithms (**QD**) [6,22] use a similar evolutionary approach. An archive of solutions is maintained and the search aims for both diversity, i.e. illuminating the archive, and individual performance, i.e. best in each niche. We consider two variations of the **QD** configuration. **QD-NS**, removing the safety stop if an operator is in the cell, increases the likelihood of collisions. **QD-NF** does not rely on fuzzy valuation for safety artefacts, only 1 for *true* and 0 for *false*, disrupting the search by reducing the granularity of the fitness metric.

We extract two features to define the niche covered by a configuration: the highest-ranked observed hazard, and the highest-ranked observed UCA. Hazards and UCAs are ranked by their identifier in the STPA analysis. This is a compromise over more accurate features, e.g. capturing the set of observed hazards, to keep the size of the QD archive reasonable. The heuristic is set to minimise the fitness metric in each niche. This configuration aims to observe a variety of safety artefacts and conditions leading to hazards in isolation of each other.

7.4   Results

We challenge our intuition on the feasibility of our approach and its underlying assumptions. Our work focused on building the Digital Twin, and the required tooling to allow for testing for safety using the proposed approach. The results in this section thus focus on state of the art heuristics for test generation[1].

Are UCAs Necessary for Hazards to Occur? We first evaluate our hypothesis of a relation between the occurrences of UCAs and Hazards outlined in Example 1, that is the occurrence of a UCA, as captured by the safety analysis, is necessary for the occurrence of a hazard. Focus of testing effort on configurations where UCAs (or components thereof) occur would increase the likelihood of discovering latent hazards.

We consider all 6000 configurations generated during our test campaigns, with the operator, arm and assembly in random states, irrespective of the heuristic. Each configuration is run through the simulation for a full cycle, the arm traversing all its way points. Processing the simulation output classifies runs into ones where no safety situation occurred, either solely UCAs or Hazards occurs, or both artefact types occur. The distribution of runs across these categories is presented in Fig. 3.

---

[1] Each batch of 1000 runs took on average 5 h to complete on a 1.8 GHz i5 laptop.

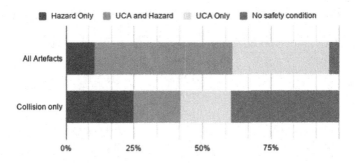

Fig. 3. Classification of situations detected across 6000 generated configurations.

We observed multiple cases where a hazard occurred without a UCA being observed as well. Such observations occurred under all heuristics. These results highlight gaps in our monitoring and safety analysis. None of the monitored UCAs in the current setup relates to the velocity of the arm, and there is no indicator that the related hazards might occur. Our initial safety analysis, at a larger scale, does capture velocity-related UCAs. However those are confined to specific, un-modelled actions of the arm, e.g. during a handover, and they still do not cover all occurrences of a hazard.

Under the "Collision only" row in Fig. 3, we focus on monitored MH-3 (a collision with the arm), and MUCA-1 (a moving arm in an obstructed cell). Our heuristics discovered solutions where the operator collided with an immobile arm (MH-3 and ¬ MUCA-1). The operator moving was not identified as a control action in our STPA analysis, given the lack of a controller on the operator in our model. As such, no related unsafe control action was identified.

We identified reasonable hazard occurrences without a related UCA. Regions where UCAs occur do result in a high likelihood of safety hazards. However, a good heuristic should consider the hazard situation components on their own to guide its result. Automated testing and classification of such occurrences helped us identify gaps in our safety analysis and monitoring, despite the explored scene and configuration space abstracting our use case.

Are Situation-Based Heuristics a Good Guide for Testing? We now consider the benefits of safety testing guided by situation-based metrics, in particular the occurrence of safety artefacts. All heuristics, except for **Ran**, use some form of feedback to identify the best configurations, and direct the search to maximise or minimise such occurrences. We compare in Table 2 the configurations generated by each heuristic, their artefacts and components coverage, and the combinations of such components they encountered.

Table 2. Coverage and fitness achieved under the different search heuristics

Heuristic	Artefact	Component (1)	Component (2)	(Min, Max) fitness	# Niches
Ran	100%	75%	54.82%	(1, 32)	6
GA-min	100%	75%	54.10%	(1, 31)	6
GA-max	100%	75%	54.69%	(1, 32)	6
QD	100%	75%	54.82%	(1, 32)	6
QD-NS	100%	75%	54.96%	(14.64, 32)	6
QD-NF	100%	75%	54.82%	(0, 31)	6

The results across heuristics are very similar. All heuristics managed to trigger all monitored artefacts in our evaluation (Artefact), cover similar combinations of them (# Niches), and the same configurations of situation components considered on their own (Comp. (1)). Considering concurrent occurrences of the safety components (Comp. (2)), the differences between heuristics remain marginal with a slight benefit for **QD**-based ones, **Ran**, then **GA**. Our simple setup has a high rate of safety occurrences, and a better comparison would be provided by a less hazard-prone setup.

No heuristic achieved a 100% coverage of the component-based coverage metrics. This is the result of gaps in the coverage, and infeasible combinations of components. As an example, three distinct components consider the position of the operator namely, in the cell, at the bench, and at the welder. Full coverage of the Comp. (2) metric would require observing the operator as being at the bench and the welder at the same time. This is infeasible under the current configuration space: the operator cannot lay down across the cell to reach both regions. Unless he climbs onto the welder, the operator will also always be considered in the cell when at the welder, $(at\_welder, in\_cell) = (True, False)$ is infeasible.

Comparing the best and worst fitness encountered by each method provides a similar insight. All methods managed to trigger all artefacts in a single configuration (a fitness of 32), although across different configurations. Evidence suggests the use of a presence sensor did reduce the likelihood of collisions between the operator and the arm, as exemplified with the high minimum fitness under **QD-NS**. Only the **QD-NF** managed to observe a safe configuration, with all configurations coming close with only a single UCA triggered.

All heuristics exploring our cage-less use-case have shown evidence of hazardous configurations. Some occurrences may be attributed to limitations of the simulated environment and setup, but they still highlight safety issues in the system. As identified in the previous section, there are few mechanisms to prevent or identify velocity constraint violations, except assessing the configuration of the arm controller, as outlined by the lack of related UCA. These could be mitigated by external control or sensing.

The proposed metrics provide little benefit to guided search heuristics, but they can help trim the configurations that need to be reviewed following an initial assessment. As an example focusing on the individuals in each of the niches discovered by **QD**, we identified a safety issue in our setup where the arm could reach outside of the cell and collide with the operator. Similarly, the arm in the confines of the cell could still collide with an operator at the bench.

8   Related Work

Combinatorial or Design of Experiments (DoE) methods [15] scope the configuration space to generate a set of complete or partial covering experiments. Gleirscher [14] suggests the use of model-based techniques to generate tests ensuring hazardous states of the system cannot be reached. Yu et al. [28] automatically derive such a model from the safety analysis results. Our approach does not rely on a model or a-priori knowledge on which situations a specific configuration triggers. We use state-of-the-art search heuristics, guided by safety-relevant metrics, to identify hazardous states.

Fontaine et al. [11] successfully propose the use of quality diversity algorithms [22] to discover failure scenarios in a human-robot shared autonomy problem. Other simulation-based approaches for testing automated vehicles show similar promising results [23]. These approaches tend to rely on metrics tailored to the modelled interaction. Our metrics instead derive from safety analyses, and fuzzy evaluation provides for a gradual progression between an unlikely and guaranteed occurrence. Neither property requires domain knowledge outside of the safety analysis.

Metrics such as MCDC [13] focus on code coverage, ensuring as an example the absence of dead code. However structural metrics are not adapted to evaluating functional requirements even more so the emergent behaviours that may arise from autonomous robots [17]. We propose instead the use of safety-centric coverage metrics, based on the observed situations, with a similar division of the decision/situation into its condition/components.

9   Conclusion

We proposed a method to assess the safety of a cobot system, relying on the artefacts captured during safety analysis to (1) inform the design of a simulation-based environment, (2) guide search heuristics towards unsafe configurations, and (3) assess confidence in the observed configurations. Our initial evaluation highlighted genuine safety issues in our setup. It shows a high hazard likelihood, and gaps in our safety analysis lead to hazards without early warning signs.

A strong safety baseline, increasing the challenge of triggering hazardous behaviours, would be required to better assess the benefits of search heuristics guided by safety components. This would also provide a fair evaluation of the proposed coverage metrics. Further refinements of said coverage metrics to trim infeasible combinations of safety components would provide improved coverage and confidence.

Acknowledgements. This project is supported by the Assuring Autonomy International Programme, a partnership between Lloyd's Register Foundation and the University of York.

We would like to thank the project partners at the University of Sheffield the University of York for numerous insightful discussions. In particular, we would like to thank James Douthwaite for his expertise with the Digital Twin.

References

1. AAIP Body of Knowledge: 1.2.1 Considering human/machine interactions. https://www.york.ac.uk/assuring-autonomy/body-of-knowledge/required-behaviour/1-2/1-2-1/cobots/. Accessed Feb 2021
2. Afzal, A., Le Goues, C., Hilton, M., Timperley, C.S.: A study on challenges of testing robotic systems. In: 2020 IEEE 13th International Conference on Software Testing, Validation and Verification (ICST), pp. 96–107. IEEE (2020)
3. Alexander, R., Hawkins, H.R., Rae, A.J.: Situation coverage-a coverage criterion for testing autonomous robots. Technical report, University of York (2015)
4. Bauer, A., Wollherr, D., Buss, M.: Human-robot collaboration: a survey. Int. J. Humanoid Robot. **5**(01), 47–66 (2008)
5. Bobka, P., Germann, T., Heyn, J.K., Gerbers, R., Dietrich, F., Dröder, K.: Simulation platform to investigate safe operation of human-robot collaboration systems. Procedia CIRP **44**, 187–192 (2016). 6th CIRP Conference on Assembly Technologies and Systems (CATS)
6. Cazenille, L.: QDpy: A python framework for quality-diversity (2018). https://gitlab.com/leo.cazenille/qdpy
7. CSI: Cobot. https://www.sheffield.ac.uk/sheffieldrobotics/about/csi-cobots
8. Dosovitskiy, A., Ros, G., Codevilla, F., López, A.M., Koltun, V.: CARLA: an open urban driving simulator. CoRR abs/1711.03938 (2017). http://arxiv.org/abs/1711.03938
9. Emerson, E.A.: Temporal and modal logic. In: Formal Models and Semantics, pp. 995–1072. Elsevier (1990)
10. Falcone, Y., Krstić, S., Reger, G., Traytel, D.: A taxonomy for classifying runtime verification tools. In: International Conference on Runtime Verification (2018)
11. Fontaine, M., Nikolaidis, S.: A quality diversity approach to automatically generating human-robot interaction scenarios in shared autonomy (2021)
12. Frigeri, A., Pasquale, L., Spoletini, P.: Fuzzy time in LTL. CoRR abs/1203.6278 (2012). http://arxiv.org/abs/1203.6278
13. Ghani, K., Clark, J.A.: Automatic test data generation for multiple condition and MCDC coverage. In: 2009 Fourth International Conference on Software Engineering Advances, pp. 152–157. IEEE (2009)
14. Gleirscher, M.: Hazard-based selection of test cases. In: Proceedings of the 6th International Workshop on Automation of Software Test, pp. 64–70 (2011)
15. Grindal, M., Offutt, J., Andler, S.F.: Combination testing strategies: a survey. Softw. Test. Verifi. Reliab. **15**(3), 167–199 (2005)
16. Guiochet, J., Machin, M., Waeselynck, H.: Safety-critical advanced robots: a survey. Robot. Auton. Syst. **94**, 43–52 (2017)
17. Helle, P., Schamai, W., Strobel, C.: Testing of autonomous systems-challenges and current state-of-the-art. In: INCOSE International Symposium, vol. 26, pp. 571–584. Wiley Online Library (2016)

18. Huck, T.P., Ledermann, C., Kröger, T.: Simulation-based testing for early safety-validation of robot systems. In: 2020 IEEE Symposium on Product Compliance Engineering-(SPCE Portland), pp. 1–6. IEEE (2020)

19. Robotics – Safety requirements for robot systems in an industrial environment – Part 1: Robots. Standard, International Organization for Standardization (2011)

20. Leveson, N.G., Thomas, J.P.: STPA Handbook, Cambridge (2018)

21. Norden, J., O'Kelly, M., Sinha, A.: Efficient black-box assessment of autonomous vehicle safety. arXiv preprint arXiv:1912.03618 (2019)

22. Pugh, J.K., Soros, L.B., Stanley, K.O.: An extended study of quality diversity algorithms. In: Proceedings of the 2016 on Genetic and Evolutionary Computation Conference Companion, GECCO 2016 Companion, pp. 19–20. ACM (2016)

23. Riedmaier, S., Ponn, T., Ludwig, D., Schick, B., Diermeyer, F.: Survey on scenario-based safety assessment of automated vehicles. IEEE Access **8**, 87456–87477 (2020)

24. Solgi, M.: geneticalgorithm: a python library for elitist genetic algorithm (2020). https://github.com/rmsolgi/geneticalgorithm

25. Vazquez-Chanlatte, M.: mvcisback/py-metric-temporal-logic: v0.1.1, January 2019. https://doi.org/10.5281/zenodo.2548862

26. Villani, V., Pini, F., Leali, F., Secchi, C.: Survey on human-robot collaboration in industrial settings: safety, intuitive interfaces and applications. Mechatronics **55**, 248–266 (2018)

27. Whitley, D.: A genetic algorithm tutorial. Stat. Comput. **4**(2), 65–85 (1994). https://doi.org/10.1007/BF00175354

28. Yu, G., Wei Xu, Z., Wei Du, J.: An approach for automated safety testing of safety-critical software system based on safety requirements. In: 2009 International Forum on Information Technology and Applications, vol. 3, pp. 166–169. IEEE (2009)

29. Zou, X., Alexander, R., McDermid, J.: Testing method for multi-UAV conflict resolution using agent-based simulation and multi-objective search. J. Aerosp. Inf. Syst. **13**(5), 191–203 (2016)

Attack and Fault Injection in Self-driving Agents on the Carla Simulator – *Experience Report*

Niccolò Piazzesi, Massimo Hong, and Andrea Ceccarelli[✉]

Department of Mathematics and Informatics, Università Degli Studi Di Firenze,
Florence, Italy
andrea.ceccarelli@unifi.it

Abstract. Machine Learning applications are acknowledged at the foundation of autonomous driving, because they are the enabling technology for most driving tasks. However, the inclusion of trained agents in automotive systems exposes the vehicle to novel attacks and faults, that can result in safety threats to the driving tasks. In this paper we report our experimental campaign on the injection of adversarial attacks and software faults in a self-driving agent running in a driving simulator. We show that adversarial attacks and faults injected in the trained agent can lead to erroneous decisions and severely jeopardize safety. The paper shows a feasible and easily-reproducible approach based on open source simulator and tools, and the results clearly motivate the need of both protective measures and extensive testing campaigns.

Keywords: Self-driving · Machine Learning · Trained agent · Adversarial attacks · Faults · Injection · Simulation

1 Introduction

Machine Learning (ML) is at the foundation of the most relevant autonomous driving applications, with more and more usage in safety critical functionalities as for example obstacle detection, lane detection, traffic sign recognition, and ultimately self-driving. This requires that the ML-based applications (and the supporting hardware) are both safe and secure i.e., faults and attacks must not jeopardize system safety [27, 28, 32]. However, the introduction of ML exposes novel attack surfaces as well as new possible fault modes.

Several works in the most recent years have investigated the effects of attacks and faults on ML-applications and have provided evidence that even small perturbations caused by hardware or software faults can deceive the ML-based application, to the extent that an incorrect output is produced [24, 25, 29]. While it is possible that a single transient fault does not lead to wrong decisions or it is masked through the activation of the different layers of the neural network [24], there is still the risk that residual software faults (when not hardware faults) manifest into an observable output corruption [26, 27, 29].

Additionally, recent works on adversarial machine learning [12] have shown that ML applications can be confused by malicious perturbations of the input i.e., by crafting

© Springer Nature Switzerland AG 2021
I. Habli et al. (Eds.): SAFECOMP 2021, LNCS 12852, pp. 210–225, 2021.
https://doi.org/10.1007/978-3-030-83903-1_14

inputs that are syntactically correct, but artificially modified such that the ML application is mistaken. Adversarial attacks could be applied as well to the inputs of a driving task [7], as we are also considering in this paper, despite they are typically though for object classification.

Consequently, solutions are required to evaluate the safety of ML applications against software and hardware faults, and to secure the system with respect to the new attack surfaces introduced by the ML applications themselves. It is not surprising that a proliferation of tools and supports to test ML-applications, including security and robustness testing, has been witnessed in recent years, altough with a general preference for classification and detection tasks [29].

Without aiming to develop new tools, but reusing existing ones, we experiment and discuss on the risk of adversarial attacks and software faults for an end2end self-driving agent, which directly maps sensory information to driving commands without organizing the self-driving pipeline in separate tasks (for example, it does not distinguish tasks in obstacle detection, localization, trajectory planning, etc.) [20]. To the best of our knowledge, no works experimented with end2end self-driving agents including both faults and attacks perspectives. Consequently, this experience report provides a guide as well as practical evidence of a method easy-to-implement that allows to rapidly deploy tests for self-driving agents under various adversarial attacks and faulty conditions, in an entirely simulated (and reproducible) environment. More in details, the paper explains how to configure, inject and ultimately collect evidences of the effects of attacks and faults injected. We show how to manipulate a driving simulator to perform the experiments and to execute the experimental campaign. Results give evidence of the effects of attacks and faults injection, especially measured in terms of wrong decisions of the trained agent that lead to collisions or traffic offences; we show that the trained agent fails under multiple injection conditions. The settings and source code we used are available at [6].

The rest of the paper is organized as follows. Section 2 reviews background notions on techniques and supporting tools that will be used in the paper. Section 3 and Sect. 4 describe respectively how we performed adversarial attacks and fault injection in the self-driving agent. Section 5 presents the experimental campaign and results. Section 6 discusses limitations of the approach and it contextualizes our work with respect to the current scenarios of testing self-driving agents against faults and attacks. Finally, Sect. 7 concludes the paper.

2 Background

2.1 Adversarial Attacks Using the ART Toolbox

Machine Learning models are vulnerable to adversarial examples, which are inputs (images, texts, tabular data, etc.) deliberately modified, while being perceptually indistinguishable, to produce a desired response by the model [17]. By adding small perturbations to original images, adversarial attacks can deceive a target model to produce completely wrong predictions [7]. In general, adversarial attacks are organized in three categories: evasion, poisoning, and extraction attacks, that we review below.

Evasion attacks modify the input to a model, typically a classifier, such that its prediction is erroneous, while keeping the modification as small as possible. Evasion attacks can be black-box or white-box: in the white-box case, the attacker has full access to the architecture and parameters of the model, and exploits this information to construct the adversarial image. In case of black-box attacks, the attacker does not have knowledge on the internals of the model; the attacker usually needs many more tries (and computational time or resources) to construct an effective adversarial image.

In *poisoning attacks*, attackers have the opportunity of manipulating the training data to significantly decrease the overall performance, cause targeted misclassification or bad behavior, and insert backdoors and neural trojans.

Last, *extraction attacks* aim to develop a new model, starting from a proprietary black-box model, that emulate the behavior of the original model.

In this work, we are relying only on evasion attacks, which have the likelihood to be carried out on a self-driving agent while it is executing, thus compromising safety. We consider both white-box and black-box attacks. White box attacks need internal details of the target model: when these details are available, they certainly pose a bigger threat. However, often the model layout can not be easily obtained. In these situations, black box attacks prove to be much more flexible, requiring only the final output of the decision process. A different approach can be to use an extraction attack to approximate the unknown target model and base a white box attack on this model instead of the original.

We do not consider poisoning attacks because they operate during the training phase of an agent: this type of attacks was not relevant to our objectives, because we are interested in investigating a trained model that runs on a vehicle. Extraction attacks are instead discarded as we are interested in safe operations rather than secrecy theft, and for the purpose of our work it is sufficient to consider the evasion attacks as explained above.

The Adversarial Robustness Toolbox (ART, [8]) is a Python library originally developed by IBM, and recently donated to the Linux Foundation. It provides the tools to craft adversarial attacks (as well as to build defenses against them). A large set of attacks from the state of the art are implemented in ART, and they can be invoked just providing as input the trained model and other parameters as the loss and optimization functions and the size of the input images. ART supports the most known Machine Learning libraries as PyTorch and TensorFlow, and it is released with the MIT open source license.

2.2 Fault Injection in Trained Agents with PytorchFI

As software fault model, we consider any software fault whose effect is modifying the value of weights or neurons in convolutional operations of the neural network during its execution. In this paper we use the tool PyTorchFi developed by Mahmoud et al. [24, 31] to modify the neuron or weights of the neural network, such that we can observe the consequences that the perturbations bring to the vehicle behaviour.

PyTorchFi is a runtime perturbation tool for deep neural networks, implemented for the PyTorch deep learning platform. It enables users to perform perturbation on weight or neurons of DNNs at runtime. PyTorchFi provides an easy-to-use API and an extensible interface, enabling users to choose from various perturbation models (or design their own custom models) [24].

PyTorchFi offers different default perturbation models that a user can select. In general, the steps to use its API are as follows: i) choose the error model, ii) specify the injection location; there can be either a single or multiple locations to perform multiple perturbations across the neural network (injection locations are specified by the layer, feature map, and neuron's coordinate position in the tensor); iii) specify whether to have the same perturbation across all elements in a batch (a batch is the number of sample utilized in one iteration: it is set to 1 in the self-driving application we use, because it processes one image at the time), or a different perturbation per element; iv) perform the injection [24].

2.3   Carla Simulator and Learning by Cheating (LbC)

The Open Urban Driving Simulator Carla (Car Learning to Act, [1]) is a simulator for autonomous driving that has been implemented as an open-source layer over the Unreal Engine 4 (UE4, [2]). Its aim is to support training, prototyping, and validation of autonomous driving models, including both perception and control. Carla includes urban layouts, several vehicle models, buildings, pedestrians, street signs, etc. The simulation platform supports flexible setup of sensor suites, and in particular we will exploit the camera sensor, that allows acquiring images from the frontal camera of a specific vehicle at a specified Frame Per Second (FPS) rate. Carla provides detailed information, collected at sampling intervals, on such vehicle while it is roving in a city. Information that can be collected is for example position, orientation, speed, acceleration, collisions and traffic violations. Further, weather conditions and time of day can be specified.

Amongst the various ML-based applications that exist for Carla, in our work we prefer a self-driving agent over other agents e.g., object recognition agents. In fact, a self-driving agent allows showing the effect of persistent faults or continuous attacks applied on consequential images, rather than on individual images without a continuous context [4]. Amongst self-driving agents, we select the trained agent Learning by Cheating (LbC) developed by Chen et al. [3]. LbC is an end2end learning [20] approach for self-driving, which directly map sensory information to steering commands. The LbC model is organized in a ResNet-34 backbone pretrained on ImageNet and three up-convolutional (upsampling) layers. LbC has demonstrated very good performance, with a minimal number of collisions under most of the environmental and traffic conditions [3].

```
while env.tick(): #at each simulation step
    observations = env.get_observations() #collect RGB image and current speed
    control= agent.run_step(observations) #compute throttle, steer, brake
    diagnostic = env.apply_control(control) #apply computed throttle, steer, brake
```

Listing 1. The main decision loop in LbC [3], [30].

The main control loop of LbC is explained in Listing 1 [30]. Using the camera as its unique input sensor, at each simulation step it is acquired (through *env.get_observations* in Listing 1): i) one *RGB* (Red Green Blue) *image* from the frontal camera of the vehicle at a resolution of 384×160 pixels, and ii) the current *speed* from the speed sensor.

These values are processed by the trained agent (*agent.run_step* in Listing 1), together with a high-level *command* ("follow lane", "turn left", "turn right", "go straight") that describes the planned route. In this way, the trained agent predicts waypoints in the camera coordinates, and then, these waypoints are projected into the vehicle's coordinate image (essentially, a trajectory is designed). From this, a low-level controller is executed (*env.apply_control* in Listing 1), that decides the steering angle, the throttle level, and the braking force. Finally, throttle, speed and braking are applied on the vehicle 3. Additional details on the LbC trained agent are outside the scope of this paper and are in [3, 30].

3 Injection of Attacks in a Self-driving Agent

We describe how we inject adversarial attacks in the self-driving agent LbC.

3.1 Selection of Suitable Attacks

We select 4 evasion attacks that we think are interesting to inject: Spatial Transformation [10], HopSkipJump [11], Basic Iterative Method [12] and NewtonFool [13]. The four attacks are chosen because they represent different approaches to the same problem, being two two black-box (Spatial Transformation, HopSkipJump) and two white-box (Basic Iterative Method, NewtonFool) approaches. The configurations we use are shown in Table 1; noteworthy, each individual attack has modifiable parameters to tweak in order to be more effective against the target model or have a more efficient computation of the adversarial example. Different combinations of parameters where tried. We ultimately decided for the values in Table 1, because in our simulation they had the better tradeoff between the effectiveness of the attack and the computational cost to generate the adversarial examples. The complete description of the attacks parameters is accessible in the official documentation [14].

Table 1. Configurations used for each attack. A detailed description of the meaning of each parameter is available in [14].

Attack	Configuration
Spatial transformation	max shift = 80%, number of shifts = 1, max rotation = 160°, number of rotations = 1
HopSkipJump	max iterations = 10, max_eval = 1000, init_eval = 100, init_size = 100, norm = 2
NewtonFool	max iterations = 10, $\eta = 0.01$
Basic Iterative Method	$\varepsilon = 0.2$, ε _step = 0.1, max_iter = 20

Spatial Transformation: The objective is to find the minimum spatial transformation that causes misclassification of an RGB image. The image is rotated by a θ angle and shifted of $(\delta u, \delta v)$ pxels. The shift is calculated as a percentage of the image size. It is a black-box attack, needing only the class prediction for the input image.

HopSkipJump: It starts from a big image perturbation and aims to reduce it to a minimum that still causes misclassification. Such image perturbation is reduced by iterations of binary searches. Each iteration produces a new perturbation, smaller than the previous one, and it stops when the boundary between the target class and the original class is reached. The distance from the original input is computed by using a norm. It is a black-box attack, needing only the class prediction for the image.

NewtonFool: It is a gradient-descent based algorithm that aims to find the perturbation that minimizes the probability of the original class. It is built under the assumption that, "nearby" the original data point, there is another point where the confidence probability in the "correct class" is significantly lower [13]. The tuning parameter η determines how aggressively the gradient descent attempts to minimize the probability of the original class. It is a white-box attack, because the attacker requires the output of the softmax function which assigns decimal probabilities to each class in a multi-class problem.

Basic Iterative Method. It is an iterative version of the Fast Gradient Method (FGM, [15]), which produces an adversarial example by calculating the perturbation that maximizes the loss (with respect to the loss of the input image). The Basic Iterative Method extends the FGM attack by applying it multiple times with small step sizes. Each intermediate result is cropped to ensure that it stays within the limits established by a hyper parameter ε, that sets the amount of perturbation allowed in the target image. It is a white-box attack because it accesses the model to compute the loss at each step.

Other relevant evasion attacks, that we left out for this work, are based on introducing modifications on the environment, rather than on the acquired images. An example is the Adversarial Patch [16] attack, which is based on crafting patches that can be applied on (or next to) objects, such that the trained agent is confused. This category of attacks was ultimately discarded, because they required a completely different approach that includes modifying assets of the simulations with the help of the Unreal Engine 4 editor.

3.2 Integration of ART in Learning by Cheating

To effectively use ART with LbC, we first define an attack module. This module contains two main functions: *load_model* and *load_attack*. The approach is the same for the four attacks.

Function *load_mode l*(*model_path*) loads the weights of the pretrained agent. It takes a file path as input. This path must point to two files: a *config.json* and the PyTorch model.*pth* containing the model weights. Clearly, the PyTorch model is the trained agent LbC, and the *config.json* is created with the model parameters. The *load_model* function returns an ART *PyTorchClassifier*, a wrapper class that contains model weights. This wrapper class allows the interaction between the attacks and the trained agent developed in PyTorch; in other words, it allows executing the successive *load_attack* on the target model.

Function *load_attack* (*classifier*, *attack*) selects one of the attacks. It requires two parameters: an instance of an ART *PyTorchClassifier* (generated at the previous step) and a string which identifies the attack. The function returns an instance of the class associated with the attack.

The injection of the attacks is done by directly modifying the source code of LbC. The trained agent is implemented in the *ImageAgent* class of LbC [3, 30]. This class is modified by adding two extra fields: *self.adv* and *self.attack*. These fields contain the targeted model and the chosen attack, loaded by calling the two functions described before.

To execute the attacks, we have also to modify the main decision loop of LbC, that is used to generate the waypoints on the basis of the collected observations. More specifically, we modify the *run_step* method from Listing 1 that generates the waypoints. In fact, the *run_step* takes as input observations the RGB image, the speed of the vehicle and the supervisor command, and feed them to the neural network to generate the next waypoints. In our version, the RGB image is modified before being passed to the network. This is done by introducing two instructions inside the *run_step*:

$$\_rgb = self.attack.generate(rgb.cpu())$$

$$\_rgb = torch.FloatTensor(\_rgb)$$

The first function creates the adversarial example; the implementation is different for each attack. The second instruction is necessary because the network processes float tensors while the *generate* function returns an array.

To better integrate ART with LbC, we need to slightly change the network decision function. In LbC this functionality is defined in the *forward* propagation function (the forward function defines how the model is going to be run, from input to output [18]) that is implemented in the *ImagePolicyModelSS* class. It creates the waypoints and it is also used by the *self.attack.generate* function to make the adversarial image. The *self.attack.generate* function only takes one parameter (the image) and it does not have access to speed and command values. This was preventing a proper invocation of the *forward* function, crashing the entire execution. The issue is solved changing how the *forward* function behaves: in the modified agent it takes only the image as a parameter, while speed and command are acquired through global variables.

4 Injection of Faults in a Self-driving Agent

We describe how we inject in the convolutional operations of the LbC trained agent.

4.1 Perturbation Models

We describe our neuron and weight injections with PyTorchFI below. We used both neuron and weight injections in our experiments with different perturbation models: some of those are default models, others are customized.

Neuron Injection. This perturbation changes the value of neurons insides the neural network with a value specified by the user. It can be on a single location or spread across the network. To perform neuron injections, the following parameters are required by PyTorchFI [24, 31]:

1. *conv_num*: the target convolutional layer.
2. *batch*: the batch in which the fault injector should run. This is set to 1 in our case, as the trained agent uses batches of a single input image.
3. *c*: the channel number.
4. *h:* the height of the input image, 160 pixels in our case.
5. *w*: the width of the input image, 384 pixels in our case.
6. *value:* the value(s) that the injector should set within the tensor.

PyTorchFi provides multiple perturbation models to simplify the injection process for the user. We use the *random_neuron_inj* and *random_inj_per_layer*. Function *random_neuron_inj* selects a single, random neuron in the network and changes its value to a random value between a range specified by the user (the default range is $[-1, 1]$). Instead, *random_inj_per_layer* selects a single, random neuron in each convolutional layer of the network, and changes its value to a random value between a range specified by the user (the default range is again $[-1, 1]$).

Weight Injection. Weight injection has the same functionality of neuron injection, but on weights. The parameters required are the same as for neuron injection, aside the replacement of batch number with the kernel number.

Because of compatibility problems we couldn't use the utility function *random_weight_location,* which returns a tuple that represents the target location of the perturbation. Instead, we implemented a custom method that calculates the injection location. We created an alternative function *random_weight* that returns a tuple containing the values that represent the location inside the network. The output of the function is a tuple including the following elements:

1. Convolutional layer number *conv_num.*
2. kernel number *k*: this is a random value between $[0,60]$. We found this interval experimentally, through multiple simulations.
3. Channel number *c*, height *h* and weight *w* of the input image.

Finally, we defined a custom function for weight injection that is essentially a copy of the default PyTorchFI weight perturbation model, with the difference that it calls our random weight locator.

4.2 Application of PyTorchFi in LbC

We need to slightly modify the *ImagePolicyModelSS* class [30] from LbC, which is executed during the *run_step* of Listing 1, by adding the following: i) import *fault_injection* and the desired perturbation models from PyTorchFi; ii) create the PyTorchFi model *pfi_model*, which is essentially a copy of the Resnet34 backbone of LbC, where injections shall be performed; iii) perform the injection on *pfi_model*, for example simply invoking the PyTorchFI API instructions:

$$inj = random\_inj\_per\_layer(pfi\_model, min\_val = -10, max\_val = 10)$$

$$h = inj(image) \text{ #image is acquired from frontal camera}$$

5   Experiments and Results

5.1   Description of the Experimental Campaign

The experimental campaign is based on the *NoCrash* benchmark from [19], designed to test the ability of vehicles to handle complex events caused by changing traffic conditions (e.g., traffic lights) and dynamic agents in the scene. In multiple runs, a target vehicle must reach a destination position B from a starting position A before a timeout expires, and under different weather conditions. The timeout value is the time required to cover the distance from A to B at an average speed of 10 km/h as in [1, 3]. For each individual run, the success criteria is that the destination B is reached before expiration of the timeout. The failure criteria is whenever the vehicle collides or the timeout expires. We include a modification of the *NoCrash* benchmark that halts the run whenever a collision occurs [4], because in our work we prioritize safety over travelled distance.

The benchmark implementation available at [30] records and saves videos of each run (for example, two screenshots are reported in Fig. 1), together with detailed logs. These videos were analyzed to visually confirm the results reported in the log and the effectiveness of the attacks and faults injected. To measure their effects, we measure the number of runs completed (the success criteria), the number of ignored red lights, and the number of collisions.

The target town we select are the Carla towns Town01 and Town02, that are basic town layouts with all "T" junctions. Town01 is the town used for training of the learned agent LbC [3], while Town02 is only used for testing. Scenarios Town01v1 and Town02v1 includes a single right turn, while scenarios Town01v2 and Town02v2 include multiple turns and crossings (Table 2). We select different values for pedestrians and vehicles, and different weather conditions as reported in Table 2. Finally, we always use the same randomization seed so that spawning positions of vehicles and pedestrians are the same in the repeated runs. Results are bound to the settings we tested, and cannot be generalized; different traffic conditions, weathers or towns could lead to different results. For the self-driving agent under consideration, the higher number of collisions is in Town02 (Town01 was also used for training the self-driving agent); this is compatible with the results in [3].

Fig. 1. Sample frame from the videos recorded. *Left*: the car is driving correctly on the intended lane. *Right*: an accident caused by the injection of HopSkip Jump. The vehicle does not steer as much as it should, and it goes out of the road, colliding with a building.

Table 2. Configuration details on the planned runs, and the results obtained for the clean runs, with 30 repetition of each configuration. The same configurations will be used for the attacks and faults injection campaigns.

Run name	Town	Vehicles	Pedestrians	Weather	Clean runs	
					Completed	Collisions
Town01-v1	Town01	20	50	Clear Noon, Wet Noon, Hard Rain, Clear Sunset	27	2
Town02-v1	Town02	15	50		26	2
Town01-v2	Town01	20	50	Wet Sunset, Soft-Rain Sunset	28	2
Town02-v2	Town02	15	50		15	13
				Sum of runs	**96**	**19**
				Runs in timeouts	**5**	
				Red lights fail	**28 (on a total of 417)**	

The experiments are organized in two phases. In the first phase, we perform *clean runs* to produce clean data i.e., we execute the simulation runs without introducing any attack or fault. Results of the clean runs are in Table 2: in this case, most of the runs are successfully completed with a limited number of collisions, and only a few red lights are ignored. Visually, we can confirm that the driving is stable: typically, the vehicle is well-placed in the middle of its lane. The vehicle only struggled in Town02-v2, where the adverse weather may have played a part in the increase of instability. In this case, most of the failures are caused by collisions, with a limited number of timeouts. The amount of collisions is clearly not appropriate for a realistic deployment of an autonomous driving system, but we clarify that they resulted from the execution of a self-driving agent which uses a single frontal camera: this is a very penalizing condition for autonomous driving applications, which usually rely on stereo cameras or lidars. Also, we remark that the scope of our work is showing the application of a campaign which includes fault and attack injections, which can be also repeated on other driving agent.

The second phase repeats the same runs of the previous phase, but with the injection of attack as described in Sect. 3, and of faults as described in Sect. 4. Attacks and faults are injected at each simulation step during the run. Results of the second phase are described in Sect. 5.2 and Sect. 5.3.

In general, a run lasted between few seconds to a few minutes of physical time, depending mainly on the point in which collisions occur.

The simulations were executed on an Intel i9-9920X@3.50 GHz CPU, 128 GB RAM, and Nvidia Quadro RTX 5000 GPU.

Concerning temporal intrusiveness of the injection, the injection of the black-box attack HopSkipJump is the most intrusive, because of the several computations required to generate the adversarial images. It should be observed that the simulation in Carla is organized in successive simulative steps: even if the physical time between two successive steps is increased (because of the computational time of HopSkipJump attack), this has small effect on the Carla simulation and its simulated time. Temporal intrusiveness of the injection of the other attacks and the faults is very small with respect to HopSkipJump.

Finally, spatial intrusiveness (amount of memory occupied) was not a concern, because of the available resources of the hardware platform we used.

5.2   Adversarial Attacks Injection: Results

An overview of the results for the injection of adversarial attacks is in Table 3, and discussed below.

HopSkipJump (HSJ). The designated route is generally followed properly by the vehicle. However, visually checking the videos, we can observe that the trajectory becomes much more unstable. The waypoints generated constantly change direction and this causes major issues, especially in curves. As for example in Fig. 1, it may happen that the vehicle does not steer as much as it should, going off-road and leading to collisions. Most of the crashes happened with the bad weather. The instability of the waypoints also leads to an increase of ignored red lights, with the vehicle entering crossroads occupied by other vehicles. Only two runs ended by timeout.

Table 3. Overview of results for the 4 injected attacks; each configuration was repeated 30 times for each attack.

	HSJ		STA		BIM		NF	
Run name	Completed	Collisions	Completed	Collisions	Completed	Collisions	Completed	Collisions
Town01-v1	27	3	28	1	0	30	13	17
Town02-v1	15	14	26	2	0	30	7	22
Town01-v2	16	13	26	4	0	30	0	30
Town02-v2	4	26	17	11	0	30	0	30
Runs sum	**62**	**56**	**97**	**18**	**0**	**120**	**20**	**99**
Timeouts	**2**		**5**		**0**		**1**	
Red lights fail	**162/339**		**26/420**		**0**		**156/227**	

Spatial Transformation (STA). The results are comparable to the clean runs. Few red lights were ignored, and the collisions happened in similar spots. In some runs the waypoint were slightly different but the trajectory was very stable. Five runs were not completed because of a timeout. STA applies a random single shift and a random rotation to each image acquired, and this may not be as much of a disruptive change as we would expect, at least with the selected parameters. We could infer that LBC is not overly sensitive to these transformations, especially since the amount of transformation is different for each image. The other three attacks use a much more refined technique, by reasoning on the decision function itself, and probably for this reason they are more effective than STA.

Basic Iterative Method (BIM). The driving is seriously compromised. In each run, it immediately starts turning right, going off-road and colliding with the surroundings. No semaphore was even reached, and no run was completed. BIM seems to be extremely effective, causing collisions right from the start of each run.

NewtonFool (NF). The agent's driving is unreliable. A very low stability in the trajectory causes serious problems in staying inside the lane. This caused a significant increase in collisions and in red lights ignored. Only one run ended with a timeout.

The results show a general decrease in reliability of the car when it is injected with the attacks. We can see the two extreme cases in the injection of STA and BIM. While the first seems to not cause any problem to the agent, the latter is very detrimental. During the other two attacks, instead, the car still functions on a basic level, but becomes much more unstable with respect to the clean runs, especially in turns and with bad weather. This indicates that these kinds of attacks can pose a real threat to the safety of autonomous vehicles and needs to be considered during development.

5.3 Faults Injection: Results

An overview of the results for the injection of faults using PyTorchFI is in Tables 4, 5 and 6, and discussed below.

Random Neuron Injection. The more we increase the range of values, the more the vehicle becomes unstable. This can be easily observed comparing the left and right side of Table 4. On the right side of Table 4, with a range of $[-10000, 10000]$ we can see that not only the self-driving agent fails to reach the goal, but all the simulations terminate almost immediately. This can be also verified by looking at the number of traffic lights encountered.

Random Neuron Injection Per Layer. It's expected that this injection causes more problems, because it injects a fault on each layer of the neural network. This is true even if the numerical value injected is much lower than in the previous case. In Table 5 (right side), within the range of $[-100, 100]$ we already obtained similar results as the single location injection with range of $[-10000, 10000]$ (Table 4, right side).

Random Single Weight Injection. The effects of weight injection are less visible than neuron injection. This can be observed in Table 6, where we have tried to increase the range of the injected erroneous value from $[-1000, 1000]$ to $[-10000, 10000]$, but without observing a significant change in the vehicle behavior: in fact, even though we can visually verify that the trajectory is unstable, the success rate is close to the clean runs in both cases.

In our results, the neuron injection per layer has the highest impact, followed by the single random location neuron injection, while the most "ineffective" model is the single

Table 4. (Left) Single random location neuron injection with values between [−1000, 1000]. (Right) Single random location neuron injection with values between [−10000, 10000].

Town	Success rate	Collision	Ignored lights	Town	Success rate	Collision	Ignored lights
Town01-v3	4/5	1/5	5/11	Town01-v3	0/5	5/5	0/0
Town01-v4	0/5	5/5	1/6	Town01-v4	0/5	5/5	0/0
Town02-v3	0/5	5/5	3/11	Town02-v3	0/5	5/5	4/6
Town02-v4	0/5	5/5	1/9	Town02-v4	0/5	5/5	2/4

Table 5. (Left) Random neuron injection per layer with values between [−50, 50]. (Right) Random neuron injection per layer with values between [−100, 100].

Town	Success rate	Collision	Ignored lights	Town	Success rate	Collision	Ignored lights
Town01-v3	4/5	1/5	7/11	Town01-v3	0/5	5/5	2/6
Town01-v4	1/5	4/5	4/7	Town01-v4	0/5	5/5	6/6
Town02-v3	0/5	4/5	6/13	Town02-v3	0/5	5/5	4/5
Town02-v4	0/5	5/5	4/5	Town02-v4	0/5	5/5	3/5

Table 6. (Left) Random weight injection with values between [−1000, 1000]. (Right) Random weight injection with values between [−10 000, 10 000].

Town	Success rate	Collision	Ignored lights	Town	Success rate	Collision	Ignored lights
Town01-v3	5/5	0/5	0/11	Town01-v3	5/5	0/5	0/11
Town01-v4	5/5	0/5	0/11	Town01-v4	4/5	1/5	1/11
Town02-v3	4/5	0/5	3/28	Town02-v3	4/5	1/5	2/27
Town02-v4	2/5	3/5	0/23	Town02-v4	3/5	2/5	1/23

random weight injection. This is most likely because the number of weights in a neural network is immensely big, and not all of them are used to process inputs.

Last, the number of runs is significantly lower than the size of the error space. However, this number is sufficient to observe that the system is indeed affected by the injected faults, as its behavior clearly differs from when faults aren't introduced.

6   Limitations and Considerations for Real-World Scenarios

We discuss two potential limitations of our study, also at the light of related works and of the possible impact for real-world scenarios.

The first limitation concerns the *amount of simulations and the selected configurations*. Input parameters to the faults and attacks injector are essential, and different configurations would clearly lead to significantly different results. This is true for both attacks and faults injection. The parameters of the attacks in Table 1, and the configuration for the generation of faults in Sect. 5, can be significantly varied and consequently can lead to i) different alterations to the input image and ii) different decisions of the trained agent. Our experimental campaign is clearly not inclusive of all the possible configurations for faults and attacks. Nonetheless, the validity of the initial objectives of our study are still valid, which are i) give evidence that a self-driving agent can fail due to software faults affecting the trained agent and due to (classifier-oriented) adversarial attacks, and ii) show an approach to quickly study these faults and attacks in a simulated environment.

The second limitation is on the *feasibility of adversarial attacks and the fault modes* in the considered scenarios. Concerning adversarial attacks, while other authors have very recently hypothesized their criticalities for autonomous driving [5, 7, 9], it is evident that implementing these attacks requires a high level of control on the attacked system. Performing an adversarial attack on a vehicle would require at least to capture the images provided by a camera and alter them before they are further processed by the trained agent. As communication between sensors and processing units for safety-critical components in vehicle is typically cabled, this would require a severe physical hacking on the vehicle, and consequently adversarial attacks are presently very difficult to apply (a different aspects is for Adversarial Patches [16], which instead can be located along the road as demonstrated in). However, as in the recent history we have witnessed also remote hacking of vehicles [21], such attack surface should not be neglected a-priori, and we believe our work can contribute as a warning on this.

Failures instead may be the consequence of bugs in neural network software [26, 27] or accelerators faults (GPUs) [23, 25]. Several recent works have raised a warning about their dangerousness [29], overall describing the possible occurrence of software and hardware faults in a similar fashion as for other safety-critical hardware and software parts of the system. Consequently, we do not need to further motivate about the possible occurrence of faults. However, the exploration of fault modes and the relative consequences on trained agents, and in particular self-driving agents, still requires investigation, as demonstrated by the many recent works on the subject e.g., [27, 28]. Again, our work can act as i) a warning on the effect in safety-critical systems, as we show that even one persistent fault injected in just one of the network layers can jeopardize functional safety, and ii) an approach to study the impact of such faults in trained agent, relying on simulated environments.

7 Conclusions

This paper describes our experience with the injection of attacks and software faults in a self-driving agent running on the Carla simulator, entirely relying on open source tools. With respect to "traditional" software, self-driving agents have peculiarities such that they exposes new attack surfaces and fault modes. The objective of this experience report is to show how these attack surfaces and new fault modes can be exploited, making it explicit that this can violate vehicle safety.

While results are clearly restricted to the set of configurations used and to the self-driving agent in use, they show the possible detrimental effects of adversarial attacks and faults, as well as the need to protect and test autonomous systems against them.

Finally, the paper aims to show how self-driving agent can be tested, relying on existing tools and a controlled environment, through a carefully described experimental campaign. Overall, our injection activities required just few modifications to the trained agent (mostly related to the forward function), and minimal knowledge on the underlying simulator.

Acknowledgment. This work has been partially supported by the project POR-CREO SPACE "Smart PAssenger CEnter" funded by the Tuscany Region.

References

1. Dosovitskiy, A.: et al.: CARLA: an open urban driving simulator. In: Conference on Robot Learning, pp. 1–16 (2017)
2. Unreal Engine. www.unrealengine.com [online]
3. Chen, D., et al.: Learning by Cheating. In: Conference on Robot Learning (CoRL) (2019)
4. Secci, F., Ceccarelli, A.: On failures of RGB cameras and their effects in autonomous driving applications. In: ISSRE, pp. 13–24 (2020)
5. Kumar, K.N., et al.: Black-box adversarial attacks in autonomous vehicle technology. arXiv e-prints 2101.06092 (2021).
6. Integration of ART and LbC. https://github.com/piazzesiNiccolo/myLbc [online]
7. Deng, Y., et al.: An analysis of adversarial attacks and defenses on autonomous driving models. In: IEEE International Conference on Pervasive Computing and Communications (PerCom) (2020)
8. Nicolae, M.I., et al.: Adversarial Robustness Toolbox v1.0.0. arXiv preprint arXiv:1807.010 69v4 (2019)
9. Zablocki, É., et al.: Explainability of vision-based autonomous driving systems: review and challenges. arXiv preprint arXiv:2101.05307 (2021)
10. Engstrom, L., Tran, B., Tsipras, D., Schmidt, L., Madry, A.: Exploring the Landscape of Spatial Robustness. In: PMLR 2019 (2019)
11. Chen, J., Jordan, M.I., Wainwright, M.J.: Hopskipjumpattack: a query-efficient decision-based attack. In: IEEE Symposium on Security and Privacy (SP) (2020)
12. Kurakin, A., Goodfellow, I., Bengio, S.: Adversarial examples in the physical world. arXiv: 1607.02533 (2016)
13. Jang, U., Wu, X., Jha, S.: Objective metrics and gradient descent algorithms for adversarial examples in machine learning. In: ACSAC 2017 (2017)
14. ART documentation v1.5.1. https://adversarial-robustness-toolbox.readthedocs.io/en/latest/
15. Goodfellow, I.J., Shlens, J., Szegedy, C.: Explaining and harnessing adversarial examples. arXiv preprint arXiv:1412.6572 (2014)
16. Brown, T.B., et al.: Adversarial patch." arXiv preprint arXiv:1712.09665 (2017)
17. Kurakin, A., Goodfellow, I., Bengio, S.: Adversarial examples in the physical world. arXiv preprint. https://arxiv.org/abs/1607.02533 (2016)
18. Stevens, E., Antiga, L., Viehmann, T.: Deep Learning with PyTorch. Manning Publications Company, Shelter Island (2020)
19. Codevilla, F., et al.: Exploring the limitations of behavior cloning for autonomous driving. In: Proceedings of the IEEE/CVF International Conference on Computer Vision (2019)

20. Grigorescu, S., et al.: A survey of deep learning techniques for autonomous driving. J. Field Robot. **37**(3), 362–386 (2020)
21. Miller, C.: Lessons learned from hacking a car. IEEE Des. Test **36**, 6 (2019)
22. Ackerman, E.: Three small stickers in intersection can cause tesla autopilot to swerve into wrong lane. IEEE Spectrum (2019)
23. Condia, J., et al.: FlexGripPlus: an improved GPGPU model to support reliability analysis. Microelect. Reliab. **109**, 1–14 (2020)
24. Mahmoud, A., et al.: Pytorchfi: a runtime perturbation tool for DNNS. In: IEEE/IFIP International Conference on Dependable Systems and Networks Workshops (DSN-W) (2020)
25. Li, G., et al.: Understanding error propagation in deep learning neural network (DNN) accelerators and applications. In: International Conference for High Performance Computing, Networking, Storage and Analysis (SC) (2017)
26. Du, X., Xiaoting, G., Sui, Y.: Fault triggers in the tensorflow framework: an experience report. In: IEEE International Symposium on Software Reliability Engineering (ISSRE) (2020)
27. Jha, S., Banerjee, S., Cyriac, J., Kalbarczyk, Z.T., Iyer, R. K.: AVFI: fault Injection for autonomous vehicles. In: IEEE/IFIP International Conference on Dependable Systems and Networks Workshops (DSN-W), pp. 55–56 (2018)
28. Jha, S., et al.: Kayotee: A fault injection-based system to assess the safety and reliability of autonomous vehicles to faults and errors. arXiv preprint arXiv:1907.01024 (2019)
29. Zhang, J.M., et al.: Machine learning testing: Survey, landscapes and horizons. In: IEEE Transactions on Software Engineering (2020)
30. Chen, D.: Learning by cheating code. https://github.com/dotchen/LearningByCheating
31. Pytorchfi documentation. https://pytorchfi.github.io/core/declare-fi
32. Zoppi, T., et al.: Unsupervised anomaly detectors to detect intrusions in the current threat landscape. ACM/IMS Trans. Data Sci. **2**(2), 7 (2021)

A Framework for Automated Quality Assurance and Documentation for Pharma 4.0

Andreas Schmidt[1] , Joshua Frey[1], Daniel Hillen[1] , Jessica Horbelt[2] ,
Markus Schandar[2], Daniel Schneider[1(✉)] , and Ioannis Sorokos[1]

[1] Fraunhofer Institute for Experimental Software Engineering (IESE),
Kaiserslautern, Germany
{andreas.schmidt,joshua.frey,daniel.hillen,daniel.schneider,
ioannis.sorokos}@iese.fraunhofer.de
[2] Fraunhofer Institute for Manufacturing Engineering and Automation (IPA),
Stuttgart, Germany
{jessica.horbelt,markus.schandar}@ipa.fraunhofer.de

Abstract. The production sector is experiencing significant transformations driven by comprehensive digitalization, interconnection, and further automation advances. One sub-sector that can benefit significantly from these trends is the production of Advanced Therapy Medicinal Products (ATMPs). ATMPs show promise for treating different serious conditions, but they are very expensive—being patient tailored products whose production is a highly manual, minimally automated process. In a recent research project with an ATMP producer, we investigated how the degree of automation can be increased. It became apparent that in parallel to increasing automation across the actual production steps, quality assurance needs to be addressed in a similar way. This paper introduces a framework for automating (parts of) the quality assurance of ATMPs using two concepts: (a) digital shadows or twins and (b) assurance cases. We demonstrate its conceptual implementation along a case study for Car-T cell products used to treat certain forms of cancer.

Keywords: Industry 4.0 · Pharma 4.0 · Advanced Therapy Medicinal Products · Assurance cases · Digital twins · Digital dependability identity · Quality assurance

1 Introduction

Industry 4.0 (I4.0) is a new paradigm in the production sector that received huge attention over the last years. The term already suggests that the adoption of I4.0 implies a fourth industrial revolution, and indeed, there is significant promise in the visions that are being pursued. A key trait of the new paradigm is the comprehensive utilization of self-aware and connected technology in production facilities to enable and facilitate seamless integration into the internet of things

© Springer Nature Switzerland AG 2021
I. Habli et al. (Eds.): SAFECOMP 2021, LNCS 12852, pp. 226–239, 2021.
https://doi.org/10.1007/978-3-030-83903-1_15

and people. Further, this increases automation up to the point of autonomy and enables frequent re-configurations for the production of specifically tailored low-volume products (lot size 1).

One specific domain that can benefit significantly from the adoption of the I4.0 paradigm is the pharmaceutical production and more specifically the production of Advanced Therapy Medicinal Products (ATMPs). ATMPs are based on genes, tissues, or cells and offer groundbreaking new opportunities in the treatment of diseases such as cancer. In many cases, however, these products need to be produced in an autologous fashion, specifically for one patient. Combining this with the circumstance that the production (including quality assurance and documentation) itself is often very labour intensive, this leads to high treatment costs of, for instance, approximately a quarter million US dollars for a CAR-T cell treatment against cancer. I4.0 (in this domain also known as Pharma 4.0) can address this issue by providing higher technical integration between the production devices—thereby enabling higher levels of automation. Apart from the production process itself, quality assurance and documentation are mandatory key concerns in the production of ATMPs. In the current state of practice, these are typically based on manual acquisition and interpretation of production data across different steps of the production chain (followed by manual documentation) which, in turn, implies high cost of labour and a significant probability for errors. Here, too, the utilization of I4.0 means could be an excellent remedy in providing higher levels of automation for the aspects of quality assurance and documentation as well, thus increasing efficiency and decreasing the probability of human errors.

In this paper, we introduce a risk-based quality assurance and documentation approach based on comprehensive model-based traceability—enabling highly automated Pharma 4.0. End-to-end traceability is enabled by an assurance-case-based argumentation structure to trace from top-level quality risks down to concrete evidences provided by the devices/machines in the production chain. Benefits are higher efficiency as well as decreased potential of human error in evidence collection, checking and interpretation (manual operations become automated). In addition, pharmacovigilance (post-market surveillance) is supported by a more formal and more understandable quality documentation. Note, however, that we do not go for full automation but rather keep the human in the loop where it is most important. Thus, we automate data acquisition and reporting, but not the actual decision making. Setting the checking bounds and reviewing the reported results is still up to a domain expert (i.e. the qualified person).

The remaining paper is structured as follows: Sect. 2 gives some insight into the domain of pharmaceutical production, particularly introduces the state-of-art in quality assurance and also highlights corresponding challenges. Section 3 introduces several digital tools and means that can be of help to overcome these challenges. In Sect. 4 we introduce our quality assurance and documentation framework that has been developed in a recent project together with experts from the production domain and a producer of Car-T cell ATMPs. Section 5 concludes the paper and gives and outlook on future research activities.

2   State-of-the-Art of Quality Assurance and Documentation in the Pharmaceutical Industry

2.1   Lifecycle of Pharmaceutical Products and Good-X-Practice

The production of pharmaceutical products is governed by various guidelines, typically referred to as GxP (Good-X-Practice), e.g. Manufacturing (GMP), Automated Manufacturing (GAMP), or Pharmacovigilance (GPvP). For *Advanced Therapy Medicinal Products* (ATMPs), the specific European GMP guidelines [3] apply and must be followed by Quality Assurance (QA). As most of these products are produced largely manually, little guidance can be found on aspects of automation (apart from digital signature processes). Furthermore, ATMPs are produced at small-scale, as these products are in fact personal, lot-size-1 products. The lifecycle of a pharmaceutical product is coarsely divided into two phases:

- Pre-Market Authorization, Research and Development
- Production of Products with Market Authorization including Post-Market Surveillance/Pharmacovigilance

In both phases, we see different QA activities that are executed either fully manually or using independent, non-integrated digital tools—hence providing potential for gains in both product safety and production efficiency for the manufacturers. Even in use cases where a hospital exemption for a single product is given, the framework we present in this paper can help to improve the quality and safety assurance process using model-based approaches.

Overall, the digitalization of pharmaceutical manufacturing is considered to be key to efficient and sustainable manufacturing, and it becomes even more important, where smaller manufacturing batch sizes require a greater need for flexibility in manufacturing plants—leading to personalized medicine. In line, digital maturity of a manufacturing plant advances through the different levels of maturity from a predominately manual and paper-based facility to a fully automated, adaptive, collaborative, self-optimizing, autonomous plant that is also fully integrated to the end-to-end, internal and external value chain (cf. Digital Maturity Model [2]). With this, Batch records develop from semi-electronic or "paper on glass" to full Electronic Batch records with review by exception. More mature status, however, includes proactive analytics across plant and internal value chain as well as integrated real-time process analytics.

2.2   The Patient Batch Record

Irrespective of the lifecycle phase, an important QA artefact is the patient batch record (also called production protocol), where the execution of steps as well as associated quality tests are acknowledged and detailed (e.g. applied process parameters). A batch record demonstrates that an organization or manufacturer properly handles and records all critical steps and actions to produce a single batch of a product, whether entered automatically or manually [15]. An electronic batch record (EBR) thereby ensures compliance and improves efficiency by automating paper-based systems.

As of today, this record is—in the best of cases—a digital free-form document (e.g. a Word document) that is filled by a production staff member, and afterwards processed by up to five other roles (lead of production, quality control lead and staff, as well as QA staff member and qualified person) for checking completeness and validity. The current form and practice of filling a batch record poses several hazards: (a) The traceability between risks (insufficient product safety) and actual countermeasures (quality tests) is provided only *manually* and *implicitly* (for definition of these terms, see [8]). (b) Entering and checking of data is not or insufficiently supported by digital tools, e.g. the transfer of data between machines is done manually as is the checking of test results as pass/fail. In summary, the batch record contains expert knowledge, but the knowledge is only implicit, e.g. the link from a certain quality test result to a potential product-induced hazard cannot be assessed in an automated way.

2.3   Pre-Market Authorization, Research and Development

Starting from a *mechanism of action* (MoA), an ATMP manufacturer develops the product as well as an associated production process, including QA. The product must be safe and efficient, while the process must be efficient and reliable.

At this experimental stage, many process steps and their parameters are not yet fixed and undergo continuous review and adaptation. As of today, a risk-based approach that starts with the hazards associated with the potential product faults and their effects on patients is not yet state-of-the-practice or not yet possible due to a lack of suitable pharmacological models. However, *model-informed drug discovery and development* (MID3) is a promising area of research [10,11] and it is expected that future regulations put a stronger focus on digital twins [21]. The associated quality assurance is also conducted in a conservative way, i.e. as much data as possible is gathered from both the production process itself as well as the therapy. In practice, this means that beside quantitative biological parameters, such as cell count and viability, also other interesting biological features are analyzed, without knowing whether a certain test result is characteristic for a safe or unsafe product.

While this might sound careless at first sight, the circumstances of patients that require treatment with ATMPs deem such an approach necessary—under certain circumstances the risks associated with non-treatment outweigh the risk associated with treating the patient with out-of-specification (OOS) product batches. At this stage, the manufacturer is in close contact with the regulation authorities to ensure rules are followed. However, these authorities cannot prescribe specific tests or QA activities as the the expertise lies with the manufacturer, as it is a novel MoA and product. Nevertheless, many manufacturers follow the ICH guidelines that describe several methods that have been validated and approved by the authorities. When making progress towards market authorization, process details settle as do the implemented test procedures. The argumentation, describing whether this is compliant with the regulations, is maintained in the form of submission documents and production protocols written in natural language—explicit, automated tracing between agreed-upon procedures and the practice is lacking.

In summary, the R&D phase requires a significant portion of expert knowledge and empirical experience gathering. Hence, an adoptable automation solution for both production as well as quality assurance steps must account for this high degree of flexibility.

2.4   Production of Products with Market Authorization

As soon as market authorization is acquired and a product specification is fixed, the employed batch record stays the same and is the central document for QA. At this stage, the major risks can be attributed to two distinct categories: (a) errors in execution, e.g. due to fatigue or human error, and (b) faults in the specification, e.g. due to an unexpected interaction between a product feature and a patient. On the business side, as the process is fixed, a non-automated approach to QA is expensive as skilled personnel, working under exceptional conditions (i.e. a cleanroom) execute tedious, automatable tasks.

When a product is used in the market, regulation demands that sufficient post-market surveillance (pharmacovigilance) is implemented. As ATMP therapies are individual and rarely applied, it is likely that the product, and in turn the manufacturer, is blamed in case of an unfavourable outcome—in contrast to other therapy products where such a causation assumption is not straightforward. In this case, the ATMP manufacturer must prove that the individual product in question has been produced in accordance with the specification co-developed with the regulation authorities. If such proof is not sufficient on the basis of process compliance (e.g. because the administration of the therapy was also correct), a root-cause analysis should be executed to remove the fault in the specification and avoid the production or usage of more unsafe products. Depending on the product, it could be that at the time of fault-detection, more products in circulation share the identified defects and must be withdrawn immediately–however with personalized products this might not often be of practical relevance.

In today's practice of manufacturing, fulfilling this task is hard as the batch records (a) are usually not accessible in digital form to, e.g., search for similar produced batches; (b) have been created in a manual fashion, leaving the residual risk of human error; and (c) lack *explicit, model-based* traceability (cf. [8]) to hazards associated with mechanisms of action and product design. The latter is due to the fact that, as of today, this knowledge is hidden in submission documents filed with the authorities and written in natural language. Hence, a specification fault is hard to detect as safety argumentation gaps, e.g. missing test procedures or unattributed physiological effects, cannot be found *automatically*.

3 Tools for Model-Based Quality Assurance and Dependability

3.1 Assurance Cases and the Digital Dependability Identity

As established in the previous section, the production of ATMPs is a safety-critical process. However, an ATMP must not only be safe but also effective, and therefore it is required to prove that several quality claims are fulfilled. An assurance case can be employed to structure a clear, comprehensive and convincing argument that the ATMP acceptably meets the level of quality for its given application by complying with relevant regulations, such as the GMP guidelines mentioned previously. Such an assurance case consists of hierarchically structured claims which are supported by evidence. One way to visualize an assurance case is by using the Goal Structuring Notation (GSN) [14], where it is structured in the form of a tree. The top goal is the root of the assurance case, typically representing fundamental claims e.g. overall safety and quality of the ATMP. Several layers of intermediate goals eventually lead to the leaf-level evidence, representing concrete facts or documentation that support their associated claims e.g. records of equipment inspection supporting claims of equipment qualification. The explicit structure of assurance cases renders them convenient for systematic review and improvement; lapses in documentation can thus be found more easily and addressed. As an example of how such a systematic review can be performed, see [6].

Today, assurance cases are usually non-digital and non-machine-readable artifacts, managed manually. For instance, evidence is documented and linked manually and changes of the system or process require a manual change of the corresponding parts of the assurance case. This approach is error-prone and scales poorly as production processes change and grow in scale and complexity. In case of an ATMP production facility, the production process can change frequently, e.g. by tailoring towards a specific patient or integrating new production pipelines. Instead of manually changing the assurance case, a digital assurance case pattern can be created. A pattern abstracts details of the case, such that it can be re-used across different applications and/or production process configurations, even throughout several stages of the same production process. We advocate the use of digital assurance case patterns that can be more easily managed and instantiated, i.e. replace their abstract parts with concrete ones, based on the specific ATMP production process at the time of instantiation. By using such patterns, assurance can be orchestrated via tool support to automatically validate supplied evidence and thus guarantee the quality of the product. Other researchers proposed an approach, where assurance cases are used to guarantee safety for the cooperation of medical devices [20]. Different devices, also from different vendors, can be integrated to operate cooperatively through defined interfaces. The safety of the cooperation is then guaranteed through the assurance case. In our advocated approach, a digital assurance case is used to systematically argue the quality in an ATMP production facility. This assurance case combines evidence and automatically evaluates the quality of the produced ATMP product. This approach also aligns with Digital Dependability Identities (DDI).

DDIs [12,13] enable seamless integration of cyber-physical systems throughout their life-cycle i.e. both during development, as well as during deployment and operation. DDIs encapsulate dependability properties (i.e. safety, security, etc.) of their subject system while offering modularity, composability, and executability. DDIs can embed assurance cases of their subject system within them, as well as the evidence relationships between the system assurance evidence, the assurance case structure, and the overall system properties that are being assured. By encapsulating all of the above information in a structured form, the DDI can be seamlessly consumed by heterogeneous tools and systems, and allows dynamic adaptation to a production system's dependability profile.

In this paper, our proposed approach exploits DDIs to create and manage digital batch records. This is achieved by deriving Digital Twins (Asset Administration Shells, see Sect. 3.2) of the batch record within an Industry 4.0 ATMP production facility. Customization needed by the ATMP pipeline can be conveniently reflected in the assurance case. Changes in the production pipeline are then captured by the DDI and the corresponding assurance case is updated.

3.2 Asset Administration Shells and Submodels

The industrial manufacturing domain is facing the digital transformation—the end-result often called Industry 4.0. While the usage of digital models (mathematical or semantic representations used for engineering) is commonplace today, the next step is to make the integration between physical world and the digital representation tighter [7,16]. Bridging this gap is done by implementing digital shadows (digital representations that "follow" the physical world) or even digital twins (digital representations that can be interacted with and behave as or interact with the physical world).

A promising approach to implementing all three forms of digital representation are the *Asset Administration Shells* (AAS) [1,19]. The AAS comes with one or more *Submodels*—well-defined collections of data as well as interactions. A Submodel is used to capture a specific aspect, use case, or in general data and operations that are relevant in a certain domain and often used together. For instance, in the discrete manufacturing sector, one Submodel can be concerned with the parameters of a drilling machine, while another carries QA data (e.g. the date and validity of the last calibration). The recent standardisation of the *Nameplate* [4] Submodel is testament to the importance of universally accessible and human- and machine-comprehensible information structures.

A major goal of the AAS and Submodels is to establish common, technology-agnostic interfaces that can be used by different vendors involved in designing, implementing, and operating automated production facilities. These interfaces carry semantic information (i.e. by referencing existing ontologies or other concept descriptions) but do not share internal implementation details. So the Submodels help to preserve intellectual property, but expose it in an interoperable way that is not a differentiating aspect of devices between vendors.

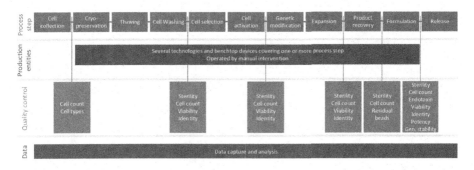

Fig. 1. A concept for an autologous CAR-T production process defines processing modules after which samples are taken and tested for their quality. Process steps, intermediate, and release test results must be documented in the batch record.

4 Framework for End-to-End Automated Quality Assurance and Documentation

4.1 A Car-T Production Process

In order to showcase this conceptual approach, we take a closer look at a concrete ATMP, namely Car-T Cells [9], a therapy product used to treat cancer. The production process is depicted in Fig. 1 and is composed of several distinct process steps. Autologous T cells are obtained from patient's apheresis material, which is cryopreserved for transport to the manufacturer. The transport of these products can only take place under strict security conditions. Subsequently, the T cells are purified after thawing and washing steps to remove the toxic freezing medium. Depending on the process, purification is performed in combination with activation of the cells using antibody-coupled beads, some of which must be removed at the end of the process. Genetic manipulation is achieved via viral approaches (lentiviral, retroviral). After expansion, cells are harvested, formulated and cryopreserved. Return transport and storage until release of the product completes the production process [17,18]. As of today, most steps are performed manually and for distinct process steps, manufacturing is enabled by the use of semi-automated and automated benchtop devices, covering one or more process steps. Still, individual devices form separated islands, that are not embedded in a centralized control software or manufacturing execution system. Additionally, the batch record has to be completed in parallel, to document all production and quality assurance activities.

Now independent of whether the production itself is automated or semi-automated, we assume that we have (at least) the following assets and associated AAS for our production system: (a) Patient Batch Record, (b) Production Devices (e.g. freezers or incubators), and (c) Quality Testing Devices (e.g. cell count or sterility tests). In the long term, there could be more assets modelled, for instance the staff members and their qualifications, but for this paper we focus on those previously mentioned.

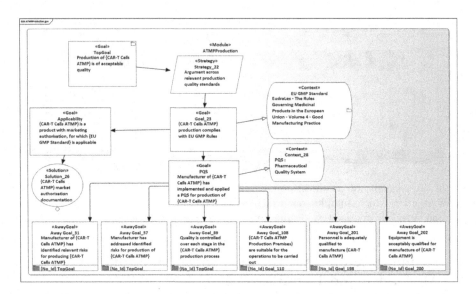

Fig. 2. High-level fragment of the ATMP assurance case, arguing for the quality of the product.

4.2 A Car-T Cell Assurance Case

The assurance case is established by claiming adherence to relevant regulations for the manufacture and administration of ATMPs. For demonstration purposes, the assurance case has been structured to claim compliance with EU guidelines for GMP of ATMPs [3], assuming an ATMP which has already received market authorization. Note that the framework can be applied—adjusting accordingly— for ATMPs without such authorization as well. In Fig. 2, the high-level argument of compliance with the guidelines is established, and then decomposed over different aspects of the latter.

In practice, the approach can adjust the assurance case structure to accommodate alternative rationales arguing compliance with different guidelines as well. The scope of the assurance case can encompass the ATMP quality starting from the manufacturing process, up to the ATMP's administration to the patient. However, for demonstrating the benefits of the approach, the upcoming discussion focuses on the process and device qualification of the overall process. Other aspects, e.g. premises qualification, can be addressed similarly.

In Fig. 3, the `TopGoal` claims that Quality is controlled across the ATMP production; the individual stages are conveniently referenced in `Context_65` nearby. Production stages are performed in sequence, as explained in `Context_66`, whereas quality is controlled in parallel via sampling (`Context_67`).

To support the above claim, a pattern over each production stage is repeated; quality for each stage is claimed based on the effectiveness of the applied controls, as confirmed by the results of the sampling. For instance, the specific argument for the quality of the cell collection stage can be seen in Fig. 4. In the figure,

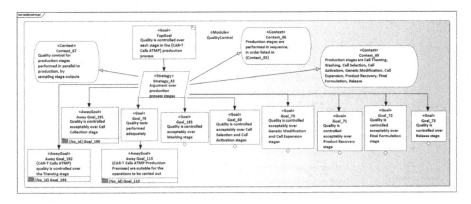

Fig. 3. Quality Control fragment of ATMP Assurance Case, arguing over the different production process stages.

the results of testing indicate acceptable quality (`Goal_76`), based on the cell count and composition results (`Goal_83` and `Goal_87`). Each test result validates the actual results from the testing (`Solution_77`), by comparing them with the acceptable lower and upper bounds (`Solution_86`). Supporting claims (`Goal_74` and `Goal_194`) argue that the chosen testing is adequate for controlling quality, and that the control devices have been appropriately qualified.

As indicated by the figures above, prior to manufacturing, the assurance case can provide placeholders for input intended to be collected during manufacturing e.g. the results of the quality tests for specific production stages. Such placeholder elements can be automatically extracted and transformed into AAS *Submodels*, which can then be used for collecting the corresponding information from quality control devices. Additional directives can be even specified e.g. to abort the production process if control ranges are violated, or to repeat sampling if needed etc. Upon successful completion of production, the AAS models can be referenced by the assurance case to review whether the produced batch record supports certification.

4.3   An Auto-Generated Batch Record Submodel

From the assurance case designed in Sect. 4.2 we now derive a *Batch Record Submodel*. Afterwards, this batch record can be instantiated in an I4.0 ATMP production facility to document relevant factors during the production and QA process—assuring the safety as well as other qualities of a specific lot-size-1 ATMP. To maintain traceability, the batch record *Submodel* should be structurally equivalent to the assurance case. This equivalence is achieved by converting the hierarchy of the assurance case into nested *SubmodelsElementCollections* and *SubmodelElements*. Maintaining the assurance case hierarchy simplifies the batch record reviewing process during the V&V process and during the investigation if a production fault is suspected after an unsuccessful treatment. To document the conformity of the production to relevant guidelines and laws, it

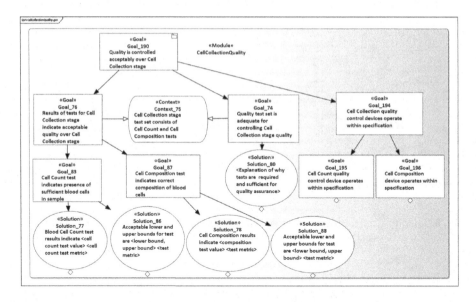

Fig. 4. Cell collection quality control fragment of the ATMP assurance case, arguing over leaf-level evidence that is provided at production time.

must be ensured that the batch record conversion is complete and correct. Therefore, a V&V process should be executed either manually or via an automated tool. In either case, the structural equivalence to the assurance case is beneficial.

Technically, the conversion from assurance case elements to a *Submodel* works as follows: An assurance case consists of tree-like nested *Goals* and *Solutions* which must be converted into corresponding *SubmodelCollections* and *SubmodelElements*. Some intermediate *Goals* are used mainly for structuring purpose and therefore they must not be documented explicitly and no further *SubmodelElements* are created within the batch record. Other *Goals* represent evidence and thus require to be documented within the batch record to approve that the safety requirements are met. In this case a *Property SubmodelElement* is created to document if a specific *Goal* is fulfilled. Another type of assurance case elements are *AwayGoals* that are converted to a *SubmodelElement* of type *Property* or *Reference*. These *SubmodelElements* are *Parameters* which are set prior to the production and refer to other authorization documents.

Further, *Goals* are specified, which correlate with the production process and must be observed and checked during runtime. As the assurance of these leaf-level *Goals* can require multiple pieces of information (e.g. a valid measurement range and a measured value placeholder), each leaf-level *Goal* is associated with an appropriate *SubmodelElementCollection*. As *SubmodelElements* can be linked to concepts from ontologies, it can be assured that, for instance, development-time safety engineering's notion of cell-counts is the same that is measured by the testing device (cf. Sect. 4.4). Additionally, the fulfillment of the *Goal*, i.e. all evidences are properly acquired, can explicitly be documented with a

Property SubmodelElement. Potentially, the batch record must check the fulfillment explicitly and thus it requires a *SubmodelElement* of type *Operation* to be specified. Effectively, the batch record *Submodel* links parts of an assurance case, as well as a specific process artifact, both of which can be derived from within a DDI. Based on the assurance case, it gathers and connects all of the required evidence during run-time by documenting relevant production and QA parameters as well as external safety evidence (e.g. certificates of a production machine).

The next step is to auto-generate the batch record *Submodel* from the assurance case. An ATMP facility can then adapt the production processes and include new processes faster and with less effort. Auto-generation can be achieved following the aforementioned process. The general structure of an assurance case for different ATMPs is going to be the identical, because all *Goals* which refer to pharmaceutical guidelines and standards are the same. Different assurance cases will then only differ in *Goals* related to the specific production and QA process activities. For auto-generation, especially these *Goals* must comply with predefined patterns and semantics to enable reliable parsing. As of today, the assurance case elements are specified in natural language. While natural language is hard to parse reliably and unambiguously, defined patterns and semantics can support a correct auto-generation of a batch record *Submodel.*

4.4   Submodels for Collaborating Assets

Our QA framework requires interaction of laboratory assets, e.g. production and quality testing devices as well as different containers and media. While the relevant data differs from process to process and machine to machine, the overall architecture stays the same: Physical assets provide information that is fed into the batch record and used to assure the product quality by comparing it to a specification, which was derived from the assurance case. In our opinion, supporting an automated QA process should be a non-differentiating factor across manufacturers and we envision that in the future this interoperability is a must and enforced by all relevant stakeholders (e.g. by industry-wide standardization). This means that apart from the batch record submodel, there are going to be submodels provided by the vendor that (a) are used to integrate it into the production line and (b) are used to integrate it into the quality assurance framework. By adopting this, the pharmaceutical production would use the same approaches as described for discrete manufacturing in [5].

An example would be a device for sterility tests—one of the final release tests in Fig. 1. This test involves adding product samples to different incubation media and then inserting them into the device. Such media are an asset whose associated Submodels contain information, such as (a) which organism types can be cultivated with it (aerobic/anaerobic bacteria, or funghi) (b) supported incubation temperature range, and (c) supported ratio between sample and medium (in weight or volume). The testing device's Submodel describes slots into which product samples with cultivation media are inserted. After testing, the device can automatically report or be queried for the binary sterility results which are directly linked to the patient batch from which the sample was taken.

5   Conclusion

In this paper, we presented a framework for significantly increasing automation of quality assurance and documentation in the pharmaceutical production domain. Applying this to the use case of Car-T cell production, we have showcased how digital tools for model-based dependability engineering (Digital Dependability Identities and embedded assurance cases) and industrial automation (Asset Administration Shells) could be used together to decrease effort and cost and at the same time maintain product quality and safety. In fact, quality and safety could even be improved, given that complex data acquisition and interpretation processes presently pose a significant challenge and are certainly error prone—a problem that is mitigated by extensive and costly manual checking practises. With this framework, safety and quality requirements as well as their fulfillment can be explicitly traced from the assurance case down to automatic checks on evidence provided by Asset Administration Shells. Process reconfiguration can also be immediately reflected by automatically updating its digital counterparts, which can then be used to generate new requirements for process validation evidence. Effectively, much of the existing effort can be alleviated, allowing ATMP development to focus resources on more critical tasks and, most importantly, to reduce product costs and thus improve accessibility to promising ATMP therapies. In future work, we want to focus on the tool-driven implementation of our approach, in particular the automatic conversion of assurance cases to Asset Shell Submodels. Furthermore, we want to investigate how model-informed drug discovery and development (MID3) can strengthen the assurance case and support the product designers in coming up with safe production parameters. In this context, we also see potential in using MID3 to support root-cause analysis.

References

1. Details of the asset administration shell - part 1. https://www.plattform-i40.de/PI40/Redaktion/DE/Downloads/Publikation/Details_of_the_Asset_Administration_Shell_Part1_V3.html
2. The development of a digital plant maturity model to aid transformation in biopharmaceutical manufacturing. https://pharmaceutical.report/whitepapers/the-development-of-a-digital-plant-maturity-model-to-aid-transformation-in-biopharmaceutical-manufacturing/1848
3. EudraLex The Rules Governing Medicinal Products in the European Union Volume 4 - Part 4 - Guidelines on Good Manufacturing Practice specific to Advanced Therapy Medicinal Products. https://ec.europa.eu/health/sites/health/files/files/eudralex/vol-4/2017_11_22_guidelines_gmp_for_atmps.pdf
4. Submodel templates of the asset administration shell - zvei digital nameplate for industrial equipment (version 1.0). https://www.plattform-i40.de/PI40/Redaktion/DE/Downloads/Publikation/Submodel_Templates-Asset_Administration_Shell-digital_nameplate.html
5. Bedenbender, H., et al.: Examples of the Asset Administration Shell for I4.0 Components. ZVEI White Paper (2017)

6. Chowdhury, T., Wassyng, A., Paige, R.F., Lawford, M.: Systematic evaluation of (safety) assurance cases. In: Casimiro, A., Ortmeier, F., Bitsch, F., Ferreira, P. (eds.) SAFECOMP 2020. LNCS, vol. 12234, pp. 18–33. Springer, Cham (2020). https://doi.org/10.1007/978-3-030-54549-9_2

7. Grangel-González, I., Halilaj, L., Coskun, G., Auer, S., Collarana, D., Hoffmeister, M.: Towards a semantic administrative shell for industry 4.0 components. In: IEEE 10th Intlernation Conference on Semantic Computing (ICSC), pp. 230–237 (2016)

8. Holtmann, J., Steghöfer, J.P., Rath, M., Schmelter, D.: Cutting through the jungle: disambiguating model-based traceability terminology. In: IEEE 28th International Requirements Engineering Conference (RE), pp. 8–19 (2020)

9. Li, Y., Huo, Y., Yu, L., Wang, J.: Quality control and nonclinical research on car-t cell products: general principles and key issues. Engineering 5(1), 122–131 (2019)

10. Marshall, S., et al.: Model-informed drug discovery and development: current industry good practice and regulatory expectations and future perspectives. CPT Pharma. Syst. Pharmacol. 8(2), 87–96 (2019)

11. Mould, D.R., Upton, R.: Basic concepts in population modeling, simulation, and model-based drug development. CPT Pharma. Syst. Pharmacol. 1(9), 1–14 (2012)

12. Reich, J., Zeller, M., Schneider, D.: Automated evidence analysis of safety arguments using digital dependability identities. In: Romanovsky, A., Troubitsyna, E., Bitsch, F. (eds.) SAFECOMP 2019. LNCS, vol. 11698, pp. 254–268. Springer, Cham (2019). https://doi.org/10.1007/978-3-030-26601-1_18

13. Schneider, D., Trapp, M., Papadopoulos, Y., Armengaud, E., Zeller, M., Höfig, K.: Wap: digital dependability identities. In: 2015 IEEE 26th International Symposium on Software Reliability Engineering (ISSRE), pp. 324–329. IEEE (2015)

14. Spriggs, J.: GSN-the Goal Structuring Notation: A Structured Approach to Presenting Arguments. Springer Science & Business Media, London (2012)

15. Stembridge, K., Adkins, M.: Making the move to electronic batch records. Pharma. Technol. 42(4), 52–55 (2018)

16. Terzimehic, T., et al.: Towards an industry 4.0 compliant control software architecture using IEC 61499 & opc ua. In: 22nd IEEE Intl. Conference on Emerging Technologies and Factory Automation (ETFA), pp. 1–4 (2017)

17. Tyagarajan, S., Spencer, T., Smith, J.: Optimizing car-t cell manufacturing processes during pivotal clinical trials. Mol. Therapy Methods Clin. Dev. 16, 136–144 (2020)

18. Vormittag, P., Gunn, R., Ghorashian, S., Veraitch, F.S.: A guide to manufacturing car t cell therapies. Curr. Opin. Biotechnol. 53, 164–181 (2018)

19. Wagner, C., et al.: The role of the industry 4.0 asset administration shell and the digital twin during the life cycle of a plant. In: 22nd IEEE International Conference on Emerging Technologies and Factory Automation (ETFA), pp. 1–8 (2017)

20. Zhang, Y., Larson, B., Hatcliff, J.: Assurance case considerations for interoperable medical systems. In: Gallina, B., Skavhaug, A., Schoitsch, E., Bitsch, F. (eds.) SAFECOMP 2018. LNCS, vol. 11094, pp. 42–48. Springer, Cham (2018). https://doi.org/10.1007/978-3-319-99229-7_5

21. Zobel-Roos, S., et al.: Accelerating biologics manufacturing by modeling or: is approval under the GBD and pat approaches demanded by authorities acceptable without a digital-twin? Processes 7(2), 94 (2019)

Fault Tolerance

A Modular Approach to Non-deterministic Dynamic Fault Trees

Sascha Müller[1]($\boxtimes$)(ID), Adeline Jordon[1](ID), Andreas Gerndt[1,2](ID),
and Thomas Noll[3](ID)

[1] Institute for Software Technology, DLR (German Aerospace Center),
38108 Braunschweig, Germany
{Sa.Mueller,Adeline.Jordon,Andreas.Gerndt}@dlr.de
[2] University of Bremen, 8334 Bremen, Germany
[3] Software Modeling and Verification Group, RWTH Aachen University,
52056 Aachen, Germany
Noll@cs.rwth-aachen.de

Abstract. Dynamic Fault Trees (DFTs) are powerful tools for deriving fault-tolerant system designs. However, deterministic approaches to DFTs suffer from semantic struggles with problems such as spare races. In this paper, we discuss the added complexity in the state-space representation of a non-deterministic DFT model and propose a modularized approach for synthesizing recovery automata. Finally, we give an implementation and evaluate it on the Fault tree FOResT (FFORT) benchmark. The results show that non-deterministic semantics with modularization can scale for literature case studies.

Keywords: FDIR · Reliability engineering · Fault Tree Analysis · Synthesis · Formal methods

1 Introduction

Radiation, limited room for human intervention under only partial knowledge, lacking the ability to replace broken hardware – space systems confront reliability engineers with many challenges. They have to ensure that spacecraft can, to a certain degree, continue operation even in the presence of faults. The ability of a system to do so is often measured by Reliability, Availability, Maintainability, and Safety (RAMS) metrics. Fault Detection, Isolation, and Recovery (FDIR) concepts aim to increase these RAMS metrics. In order to derive these concepts and evaluate them, reliability engineers employ Fault Tree Analysis (FTA) [8]. A Fault Tree (FT) is a graphical failure model describing how low-level faults propagate through a system and eventually become a system-wide failure. To strengthen the expressive power of FTs, they were later extended to Dynamic Fault Trees (DFTs), which introduce various features such as temporal dependencies and spare management. However, these DFTs give rise to non-deterministic

© Springer Nature Switzerland AG 2021
I. Habli et al. (Eds.): SAFECOMP 2021, LNCS 12852, pp. 243–257, 2021.
https://doi.org/10.1007/978-3-030-83903-1_16

behavior such as spare races. In these spare races, multiple resources compete simultaneously for a spare, but have no unique semantic resolution.

A methodology presented in [14] aims to overcome this shortcoming. It introduces Non-Deterministic Dynamic Fault Trees (NdDFTs), a non-deterministic extension of DFTs, that drops the inherently rigid rules on how spares should be employed. The methodology foresees transforming NdDFTs to Markov Automata (MA), computing an optimized scheduler for a given objective metric, and then extracting the recovery strategy from said scheduler. This process is referred to as recovery automaton synthesis.

However, as the technique constructs a monolithic state-space representation of the NdDFT, containing an encoding of all possible recovery actions, it suffers severely from classical state-space explosion problems. When employing the technique for industrial benchmarks, we experienced that its naive usage makes it unsuitable for real life applications.

The main contribution of this paper lies in tackling this weakness. For that purpose, we adapt established modularization techniques from the DFT realm. These techniques were originally designed to compute reliability metrics in a compositional way. We transfer them to establish a modular workflow for performing recovery automaton synthesis.

The remainder of the paper is structured as follows. Section 2 gives an overview of other approaches that address non-deterministic semantics in the context of DFTs. Technical background knowledge on the relevant fault tree models is given in Sect. 3. Section 4 extends the synthesis workflow to integrate modularization. Section 5 investigates the scalability of the approach. Finally, the paper concludes in Sect. 6, and gives further directions to future work.

2   Related Work

The problem cases induced by the rigid standard fault tree semantics have been considered in other works. The authors of [9] tackle the issue of spare races by employing non-determinism in the propagation semantics of functional dependency gates, while allowing only functional dependencies to cause spare races. The study of the interaction of this approach with spare gates reveals that there are various different, yet sensible, ways in which the resulting semantics can be interpreted. They conclude that there is no "correct" one-fits-all interpretation, and that the fitting variant has to be chosen on a case-by-case basis. This is a concern regarding the applicability of fault trees, as experts in system design are not necessarily experts in fault tree semantics.

The work by the authors of [2] is worthy of particular mention. They apply an approach for converting static, non-deterministic fault tree models, extendible with so-called repair boxes, into Markov decision processes. Resolving the nondeterminism gives them an optimal repair strategy. However, the approach does not extend to dynamic gates, which the authors mention as foreseeable future work.

The authors of [18] introduce the concept of fault maintenance trees, which are based on non-deterministic Input-Output Interactive Markov Chains

(I/O-IMCs). The semantics are defined compositionally by starting with elementary I/O-IMCs for every gate and by proceeding in bottom-up direction by combining these elementary I/O-IMCs. Recovery strategies for resolving spare races can also be given on the Markovian level. The strategies have to be chosen manually and are then compared to each other using a testing-based approach based on model simulation. In contrast, the NdDFT model employed in this work defines recovery strategies on a higher level and allows them to be computed by resolving the non-determinism on the Markovian level with respect to an optimality criterion.

Also with regard to modularization techniques, a number of techniques to support compositional analysis of DFTs has been developed. For identifying minimal cut sets, [5] provides an algorithm for finding independent sub-modules of FTs, which can be converted separately to Binary Decision Diagrams (BDDs) and then be analyzed, reducing the computational requirements for handling the entire tree. Our approach to modularization as described in Sect. 4.2 is based on this work.

For efficiently calculating the reliability of a DFT, [3] provides a compositional semantics for DFT in terms of IMCs, which reduces the combinatorial explosion in many common cases. Moreover, [16] develops an approach to divide a DFT into independent sub-modules for computing reliability. Sub-modules containing only static gates can then be solved using a traditional BDD method, while sub-modules containing dynamic gates can be solved using Markov Chain analysis. The method presented in [7] also modularizes a DFT and uses BDDs for the static sub-modules, but employs the approximation from [1] to solve the dynamic sub-modules. This avoids the state-space explosion problem incurred by conversion to Markov Chains, while retaining a reasonable degree of accuracy. Based on this work, [11] proposes a method to modularize DFTs further, by also collapsing static sub-trees of a dynamic gate, but keeping additional information about the probability distribution of these sub-trees. Finally, [20] provides additional modularization techniques, which can convert static sub-trees and some dynamic sub-trees into equivalent basic events, thus reducing the complexity of further analysis.

3   Background

3.1   Fault Trees

Fault trees are failure propagation models that express how faults start out on low-level components, propagate through the system by combinatorial means, and eventually turn into a high-level, system wide failure. Syntactically, fault trees are directed acyclic graphs with two types of nodes: events and gates. The leaves of a fault tree are called basic events, and the root node is referred to as top-level event. Usually, a basic event is also equipped with a failure rate. In this work, we allow for one additional extension: A basic event (BE) can be equipped either with a failure rate, for describing exponentially distributed behavior, or with a failure probability for instantaneous, uniformly distributed

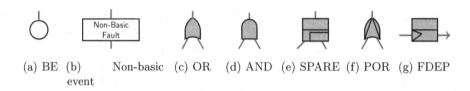

(a) BE (b)   Non-basic (c) OR   (d) AND (e) SPARE (f) POR (g) FDEP
          event

Fig. 1. Relevant gates and events of a fault tree

behavior. When a basic event with a uniform distribution is activated, it can fire with its assigned probability. Such an activation occurs at system start and may be triggered again by a gate. We limit ourselves here to the case of permanent failure, i.e., once a BE has failed, it remains in a failed state for all future points in time.

The gates model logical re-combinations of faults, as they propagate through the system. The simplest type of fault tree model, called static fault tree, considers basic logical gate types such as AND and OR. Dynamic Fault Trees (DFT) go further and introduce new gate types enabling various features such as spare management and temporal constraints.

We give a short overview of gates relevant to this paper. Their fault tree notation, together with the notation of events, is depicted in Fig. 1. The OR and AND gates behave as classical, logical gates. If at least one input fails, the OR gate propagates. If all inputs fail, the AND gate propagates.

The SPARE gate has a primary event and a set of spare events, also called the spare pool. Spare events can be shared, and are initially deactivated. However, sub-trees of spare events cannot have any shared nodes. When an input fails, the SPARE gate claims a spare in left-to-right order, and activates it. If the primary input has failed, and all available spares have failed, the SPARE gate fails.

The POR (priority OR) gate propagates if and only if the left-most input occurs before any other input.

The FDEP (functional dependency) gate has a triggering event and any number of dependent basic events. When the triggering event occurs, the dependent basic events are also set to failed. Syntactically, the triggering event and the dependent events are defined to be inputs to the FDEP gate. To prevent semantic confusion, however, the graphical representation uses outgoing edges to connect dependent events.

To illustrate the DFT notation, we consider the example shown in Fig. 2. The depicted DFT consists of four memory components; two primaries and two spares. The two spares are part of a spare pool shared among the two SPARE gates. According to standard DFT semantics, priority is given to claiming Memory3 before Memory4 in case of a failure of Memory1 or Memory2. Moreover, the system is equipped with two hot redundant, always active power sources, Power1 and Power2. Power1 powers both primary components, Memory1 and Memory2 and Power2 powers the spares. Finally, FDEP gates are used to model the functional dependencies between power supplies and memory components. The FDEPs propagate the failure of a power source to the respective memory components.

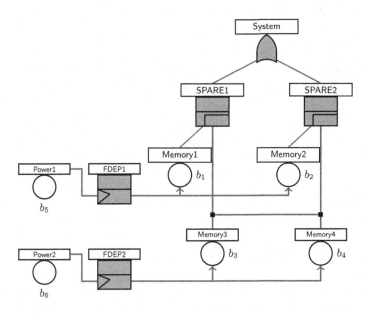

Fig. 2. Example DFT

3.2   Non-deterministic Dynamic Fault Trees

DFTs impose a fixed and rigid order in which spares are activated. They do not allow to adapt the order depending on the history of occurred faults. This may lead to semantically undesirable consequences:

- A SPARE gate might claim a spare from a spare pool, despite having an already failed parent. This might deny a necessary resource to other SPARE gates that urgently require the spare to recover.
- In the event of spare races, it is not semantically clear which SPARE gate may claim a spare.
- The optimal order for spares has to be known at design time of the fault tree.

Figure 3 visualizes possible DFT configurations exhibiting the above described semantic complications. Red indicates an incoming failure propagation. Spare claims are marked with thick, black lines.

In order to overcome these issues, [14] introduces an inherently non-deterministic DFT model (NdDFT, following the naming in [2]), which relaxes the semantic restriction of DFT models. Syntactically, the notation of the NdDFT is adopted from the DFT. Semantically, the NdDFT introduces natural non-determinism for spare activations, by allowing a SPARE gate to choose freely which spare to pick. A SPARE gate may also decide to not claim any spares and leave them available for other SPARE gates. In other words, the following recovery actions can be taken:

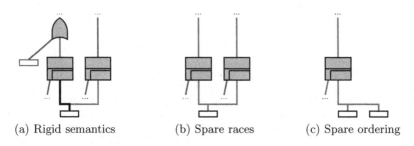

(a) Rigid semantics   (b) Spare races   (c) Spare ordering

Fig. 3. Example configurations of problematic DFTs

Definition 1 (Recovery Action). *A recovery action r in an NdDFT T is an action of the form*

- *[] (empty action) or*
- *$CLAIM(G, S)$ (spare gate G claims spare S, where S is a spare of G).*

The non-determinism in NdDFT models is then resolved via an object called Recovery Automaton (RA). The RA defines which recovery actions should be taken whenever a set of basic events occur. The reason why sets of events as opposed to single events are used, is due to the ability of FDEPs to cause several basic events to fail simultaneously. Likewise, in order to react to the simultaneous occurrence of basic events, the RA may need to perform not just a single recovery action, but a sequence of recovery actions.

To introduce the formal notion of the RA we formalize the above auxiliary concepts as follows: For that, we denote the set of all recovery actions possible in an NdDFT T by $R(T)$. Moreover, this definition is extended to the set of recovery action sequences through $RS(T) := (R(T)\backslash\{[]\})^*$. For recovery action sequences, the empty action is ignored and considered as the empty word ϵ. The * here denotes the usual Kleene closure. Similarly, we denote the set of all non-empty subsets of basic events of an NdDFT T by $BES(T)$. We now introduce the RA on a formal level.

Definition 2 (Recovery Automaton). *A Recovery Automaton (RA) $\mathcal{R}_T = (Q, \delta, q_0)$ of an NdDFT T is an automaton where*

- *Q is a finite set of states,*
- *$q_0 \in Q$ is the initial state, and*
- *$\delta: Q \times BES(T) \rightarrow Q \times RS(T)$ is a deterministic transition function that maps the current state and an observed set of faults to the successor state and a recovery action sequence.*

To illustrate the interaction between RA and NdDFT, we give a simple example in Fig. 4. The system has a cold redundant spare and according to the RA, the redundancy is activated upon failure of the primary unit.

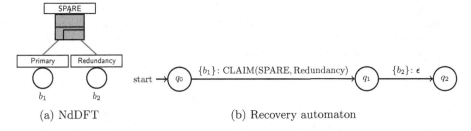

(a) NdDFT   (b) Recovery automaton

Fig. 4. Example of (a) NdDFT and (b) RA

3.3 Synthesizing Recovery Strategies

We sketch the key steps for recovery automaton synthesis. Further details are available in [14]. Initially, the algorithm converts an NdDFT model into a so-called Markov Automaton (MA) [6]. An MA is a transition system with continuous-time, non-deterministic, and probabilistic transitions. The MA contains all possible decisions on spare activations.

The transformation of an NdDFT into an MA is obtained by adapting traditional state-space generation algorithms for transforming DFTs to Continuous-Time Markov Chains (CTMCs). The base algorithm adapted here is given in [4]. The adapted algorithm operates in the following manner:

- Each state tracks a history of occurred basic event sets $(B_1, B_2, \ldots, B_n)$ and a mapping from spare gates to the currently claimed spares.
- The algorithm starts with the initial state denoted by ().
- BEs which are activated or have a dormant failure rate > 0 are considered as enabled events. Enabled events are used to compute the Markovian successors of a state. The history of the successor state is extended accordingly.
- The basic event set is obtained by taking a failing enabled event and computing the transitive closure according to FDEPs.
- Markovian transitions are labeled by the failure rate of the failing BE.
- All transitions that would lead to a state implying the top-level event are instead redirected to a special FAIL state.
- For each Markovian successor, non-deterministic successors are then computed, each of them corresponding to an enabled recovery action sequence.

The optimal recovery strategy, represented by an RA, can then be obtained by optimizing the scheduling of the generated MA with respect to an objective metric. The RA is then further reduced using both common state-space reduction methods based on trace equivalence and techniques exploiting the domain knowledge about the occurrence of faults [15]. Finally, by performing model checking queries on the Markov Chain (MC) obtained from the RA, enriched with the corresponding failure rates of the NdDFT, the desired RAMS metrics can be computed. A summary of the workflow and simple examples of all involved semantic objects can be found in Fig. 5.

The metrics computed from the MC may be the optimization objective, but might also be any other metric of interest. The RA ensures that a consistent

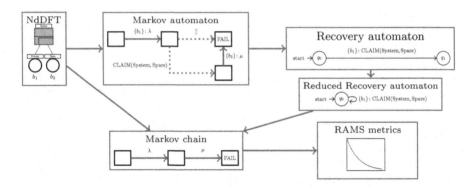

Fig. 5. Transformation road map

recovery strategy is applied for all possible queries. Relevant metrics of interest are for example:

- **Reliability After Time** t, which describes the probability that a system is still functional after a time span t.
- **Mean Time To Failure** (MTTF), which describes the expected time span that will pass until the system-level failure occurs.

The concrete metric itself is interchangeable. However, in this work, we focus on optimizing with regard to MTTF. This gives the advantage of dropping the time parameter t. For the MA, maximizing the MTTF corresponds to maximizing the expected long-term reachability property of the FAIL state.

4 Modular Synthesis of Recovery Automata

The Markovian state space generated from a fault tree can be massive. In general, its size can grow exponentially with the number of nodes in a fault tree. The problem of an exponentially increasing state space is commonly known as the *state-space explosion problem*. In conventional dynamic fault trees, the blow-up can be attributed mostly to the interleaving occurrence of basic events. In the case of non-deterministic DFTs, the state-space explosion problem gains an additional dimension: The non-determinism caused by the selection of an appropriate recovery action generates an additional source of exponential blow-up.

Ensuring scalability while synthesizing recovery strategies for large fault trees with hundreds of basic events is nearly impossible using the previous, naive workflow. In the following, we consider how existing modular approaches for deterministic DFTs can be leveraged to solve the synthesis problem.

4.1 Modular Workflow

To tackle the state-space explosion problem for calculating RAMS metrics on deterministic DFTs, previous works have considered employing modularization techniques. These primarily involve detecting independent sub-trees in a fault tree – referred to as modules –, evaluating the metrics on the individual modules, and then composing them into the total metric for the original fault tree. For example, the total reliability after a certain time span for two modules connected via an AND gate can be obtained by means of multiplication.

Commonly, this approach faces a significant issue: Not all metrics can be computed in a compositional manner, but instead require the full state space for computation. In particular, highly interesting metrics such as the MTTF are not compositional [19]. However, in the context of the recovery strategy synthesis problem, the problem of compositionality changes. Even though the metric to be optimized may not be compositional, to determine the recovery automaton it is fortunately not necessary to compute the actual metric for the complete tree. Instead, the objects that require composition are the already optimized recovery automata. Automata composition in turn is a common problem that can be solved using standard techniques. We therefore exploit a two-stage approach by first synthesizing recovery automata, and then employing them during the computation of the actual metrics. In this manner, the non-determinism can be resolved modularly during the synthesis step. In greater detail, we apply the following approach:

1. Modularization: determine the modules in the fault tree.
2. Trimming: discard modules without non-determinism.
3. Synthesis: compute the optimal RA for each module, and reduce it.
4. Composition: assemble the overall recovery automaton from the modular RA.

As noted previously, basic events are a major driver for exponential blow-up. Therefore, events that do not affect the resolution of the non-determinism are taken out of the equation. Finally, as the non-determinism has already been resolved before the evaluation step, this particular source of exponential blow-up is absent during the computation of the metrics. Trimmed modules are only discarded for the purpose of the RA synthesis. This ensures that unnecessary information is safely removed, but properly considered during the metrics computation. The new workflow incorporating modularization is visualized in Fig. 6.

4.2   Modularization

We base our modularization approach on the pre-existing algorithm given in [5]. It applies a depth-first search on the fault tree, traversing all nodes while keeping track of the first and last visiting time of each node. These visiting times are then used to identify the modules using the following criterion: Given a node which is suspected of being the root of a module, if its descendants' visit dates – both first and last – all lie within the first and last visit dates of that node, then the node and all of its descendants form a module. In addition to this basic rule, further restrictions have to be applied to obtain the desired compositionality property for the recovery automata.

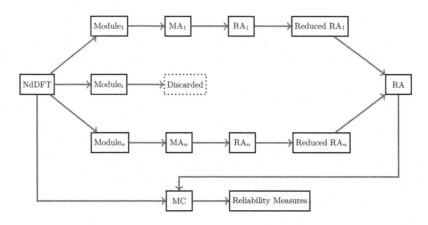

Fig. 6. Transformation road map with modularization

There are two special cases which have to be considered: SPARE gates and all types of priority gates. Priority gates are road blockers to the desired compositionality property, as they may change the optimization direction. Consider for example a POR gate. In the case of the first input being a SPARE gate, the optimal strategy for maximizing the MTTF would also be to maximize the MTTF of the SPARE gate. In other words, claiming its available spares is the best course of action. On the other hand, were a SPARE gate the second input to a POR gate, then suddenly this simple relationship changes: Now minimizing the MTTF of the SPARE gate will lead to a scenario where the POR gate is more inclined to become fail-safe. The two scenarios are visualized in Fig. 7.

Therefore, given recovery automata for two modules connected by a POR node, we cannot obtain the overall recovery automaton by means of composition. In addition to priority gates, SPARE gates also prohibit further modularization of their sub-trees. Due to the semantic definition of a SPARE gate, any basic event contained in a sub-tree may trigger a recovery action, and thus requires a representation within the Markovian state space. Bundling these observations, we obtain the following restrictions of the modularization rules:

- A SPARE gate that is a descendant of a priority gate cannot be the root of a module.
- A node that has a SPARE gate as a descendant and that is a descendant of a priority gate cannot be the root of a module.
- A descendant of a SPARE gate cannot be the root of a module.

Finally, an example application of the algorithm with the additional rules is given in Fig. 8. The algorithm proceeds in a leftmost order. Each node is labeled by the first and last visiting time, and the computed modules are indicated by dotted boxes.

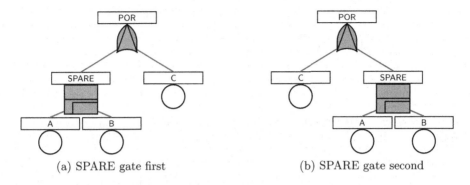

(a) SPARE gate first   (b) SPARE gate second

Fig. 7. Non-compositionality of priority gates as they change optimization direction.

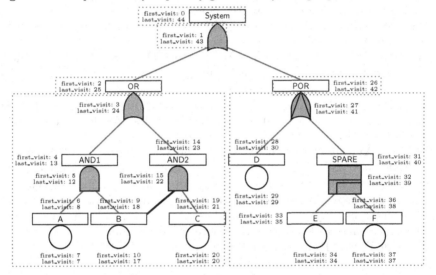

Fig. 8. Example application of the modularization algorithm

5   Case Studies

The proposed non-deterministic semantics and the modular workflow have been implemented within our application Virtual Satellite 4 FDIR (VirSat FDIR) [13]. Virtual Satellite 4 is an Eclipse-based modeling framework intended for Model-Based Systems Engineering of spacecraft [10]. VirSat FDIR is an application employing the framework to provide capabilities for modeling FDIR.

The FFORT benchmark set introduced in [17] was used as a source of fault tree benchmarks to evaluate our proposed techniques. FFORT is an online fault tree database with fault trees collected from scientific literature for the primary purpose of benchmarking. From the FFORT benchmark set, we have selected fault trees which contain at least one SPARE gate but do not employ the authors' custom fault tree extension of inspection modules (IM). Therefore, we can guarantee that all experiments contain some non-determinism. The following fault

Table 1. Summary of benchmark results

Modularization	Solved	Timeouts	OOMs	solveTime [s]
No	22/156	10	124	1429
Yes	142/156	0	14	292

tree families from the FFORT benchmark fulfilled the selection criteria. (The graphic symbols refer to the evaluation charts shown in Fig. 9 and 10.)

- **Active Heat Rejection System (AHRS ○).** The AHRS is made up of thermal rejection units of which only one is needed for the system to function.
- **Cardiac Assist System (CAS ●).** The CAS models a hypothetical cardiac assist system with redundant CPUs, motors, and pumps.
- **Electro-Mechanical Actuator (EM _).** The model focuses on common-cause failures in an electro-mechanical actuator.
- **Hypothetical Example Computer System (HECS □).** The HECS fault trees model computer systems including their processors, memory modules, buses, consoles, operators, and software.
- **Hypothetical Example Multi-Phase System (HEMPS ■).** The HEMPS model is a demonstrator of a system designed for a multi-phase mission.
- **Mission Avionics System (MAS *).** The MAS models represent mission- and safety-critical systems with high redundancy. Components include hardware, software and vehicle control subsystems, and system management.
- **Multiprocessor Computing System (MCS ×).** The MCS model computers with power supplies, memory modules, hard disks, and connecting buses. The benchmarks have been enriched with instances from [19].
- **Nuclear Power Plant Water Pumping System (NPPW ⊗).** The model represents a nuclear power plant system.
- **Railway Crossing (RC ○).** The RC fault tree collection models level railway crossings with sensors, motors, and controllers. The models come in two variations (sc and hc), representing the controller being a single basic event or hypothetical example computer system, respectively.
- **Vehicle Guidance System (VGS △).** The VGS models are industrial case studies dealing with variants of safety concepts for vehicle guidance systems.

The benchmarks were carried out with a Intel i7-6600U CPU, 4 GB of RAM, and a timeout of 600 s. The software, the experiment setup, all experiments, and all results can be found at [12]. The number of solved instances, the number of timeouts, the number of out-of-memories (OOMs), and the total solving time were logged. A summary of the results is given in Table 1.

As hypothesized in the beginning of the paper, not applying modularization leads to massive state-space explosion, causing many cases of OOMs. Applying modularization, on the other hand, yields a major speed-up, enabling us to synthesize RA for nearly all instances. An interesting observation is also that the number of timeout events is rather small, compared to the large number of

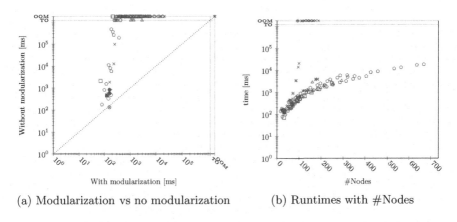

(a) Modularization vs no modularization   (b) Runtimes with #Nodes

Fig. 9. Time measurements of modularization approach

OOMs. This suggests that there is room for investing more computation time into further techniques for state-space reduction, and hence also reducing the memory consumption.

The following charts give a closer look at the results of the experiments. Figure 9a shows a detailed time comparison between the synthesizer with and without a modularizer, respectively. How the algorithm scales overall as the total number of fault-tree nodes increases, is shown in Fig. 9b. The dashed line marks where both algorithms require the same time. Timeouts and out-of-memory results have been placed on the outer lines and are labeled with TO and OOM, respectively.

As described before, two major drivers for state-space explosion are basic events and SPARE gates. Figure 10a gives a breakdown on how the number of BEs impact the synthesis. Likewise, Fig. 10b gives the breakdown in reference to the number of SPARE gates. Modularization is enabled in both cases. The data shows that the speed-up gained from modularization is overall crucial to obtain scalability, but also heavily depends on the fault tree family. The families causing OOMs are primarily MAS and MCS. A closer look into Fig. 9b and Fig. 10b reveals that these have a relatively small number of nodes, while at the same time having a relatively large number of SPARE gates.

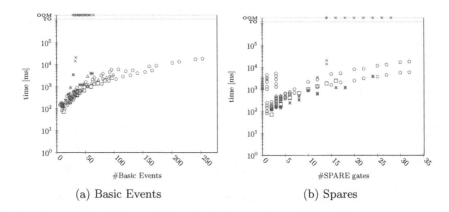

(a) Basic Events   (b) Spares

Fig. 10. Time measurement break-down for basic events and SPARE gates

6 Conclusions and Future Work

In this paper, we investigated a modular approach approach to recovery automata synthesis for non-deterministic DFTs. In order to deal with the increasing complexity due to the semantic extension, modularization approaches were employed. Both their necessity and effectiveness were demonstrated on case studies coming from the FFORT benchmark set. However, it was also shown that the semantics still yield a severe level of state-space explosion, causing many out-of-memories but only few timeouts.

Further techniques for dealing with larger modules and modules with a high degree of non-determinism are therefore desirable. In the past, symmetry reduction techniques have proved useful to combat the state-space explosion problem in deterministic DFTs [19]. In the future, we hope to investigate how those approaches can be leveraged to NdDFTs to further improve the efficiency of our approach.

References

1. Amari, S., Dill, G., Howald, E.: A new approach to solve dynamic fault trees. In: Annual Reliability and Maintainability Symposium, pp. 374–379. IEEE (2003). https://doi.org/10.1109/RAMS.2003.1182018
2. Beccuti, M., Franceschinis, G., Codetta-Raiteri, D., Haddad, S.: Computing optimal repair strategies by means of NdRFT modeling and analysis. Comput. J. **57**(12), 1870–1892 (2014). https://doi.org/10.1093/comjnl/bxt134
3. Boudali, H., Crouzen, P., Stoelinga, M.: A compositional semantics for dynamic fault trees in terms of interactive Markov chains. In: Namjoshi, K.S., Yoneda, T., Higashino, T., Okamura, Y. (eds.) ATVA 2007. LNCS, vol. 4762, pp. 441–456. Springer, Heidelberg (2007). https://doi.org/10.1007/978-3-540-75596-8_31
4. Dugan, J.B., Bavuso, S.J., Boyd, M.A.: Dynamic fault-tree models for fault-tolerant computer systems. IEEE Trans. Reliab. **41**(3), 363–377 (1992). https://doi.org/10.1109/24.159800

5. Dutuit, Y., Rauzy, A.: A linear-time algorithm to find modules of fault trees. IEEE Trans. Reliab. **45**(3), 422–425 (1996). https://doi.org/10.1109/24.537011
6. Guck, D., Hatefi, H., Hermanns, H., Katoen, J.-P., Timmer, M.: Modelling, reduction and analysis of Markov automata. In: Joshi, K., Siegle, M., Stoelinga, M., D'Argenio, P.R. (eds.) QEST 2013. LNCS, vol. 8054, pp. 55–71. Springer, Heidelberg (2013). https://doi.org/10.1007/978-3-642-40196-1_5
7. Han, W., Guo, W., Hou, Z.: Research on the method of dynamic fault tree analysis. In: International Conference on Reliability, Maintainability and Safety, pp. 950–953. IEEE (2011). https://doi.org/10.1109/ICRMS.2011.5979422
8. International Electrotechnical Commission, Geneva, Switzerland: Fault Tree Analysis (FTA) (2006)
9. Junges, S., Guck, D., Katoen, J.P., Stoelinga, M.: Uncovering dynamic fault trees. In: 2016 46th Annual IEEE/IFIP International Conference on Dependable Systems and Networks (DSN), pp. 299–310. IEEE (2016)
10. Lange, C., Grundmann, J.T., Kretzenbacher, M., Fischer, P.M.: Systematic reuse and platforming: application examples for enhancing reuse with model-based systems engineering methods in space systems development. Concurrent Eng. **26**(1), 77–92 (2018). https://doi.org/10.1177/1063293X17736358
11. Liu, D., Xiong, I., Li, Z., Iang, P., Zhang, H.: The simplificafion of CUF sequence SEF analysis for dynamic systems. In: International Conference on Computer and Automation Engineering, pp. 140–144. IEEE (2010). https://doi.org/10.1109/ICCAE.2010.5451831
12. Müller, S.: virtualsatellite/VirtualSatellite4-FDIR: Release 4.12.1, October 2020. https://doi.org/10.5281/zenodo.4075576
13. Müller, S., Gerndt, A.: Towards a conceptual data model for fault detection, isolation and recovery in Virtual Satellite. In: SECESA 2018. European Space Agency (2018). https://elib.dlr.de/122061/
14. Müller, S., Gerndt, A., Noll, T.: Synthesizing FDIR recovery strategies from non-deterministic dynamic fault trees. In: 2017 AIAA SPACE Forum, vol. AIAA 2017–5163. American Institute of Aeronautics and Astronautics (2017). https://doi.org/10.2514/6.2017-5163
15. Müller, S., Mikaelyan, L., Gerndt, A., Noll, T.: Synthesizing and optimizing FDIR recovery strategies from fault trees. Sci. Comput. Program. **196**, 102478 (2020). https://doi.org/10.1016/j.scico.2020.102478
16. Pullum, L., Dugan, J.: Fault tree models for the analysis of complex computer-based systems. In: Annual Reliability and Maintainability Symposium, pp. 200–207. IEEE (1996). https://doi.org/10.1109/RAMS.1996.500663
17. Ruijters, E., et al.: FFORT: a benchmark suite for fault tree analysis. In: 29th European Safety and Reliability Conference, pp. 878–885. Singapore Research Publishing (2019). https://doi.org/10.3850/978-981-11-2724-3_0641-cd
18. Ruijters, E., Guck, D., Drolenga, P., Stoelinga, M.: Fault maintenance trees: reliability centered maintenance via statistical model checking. In: 2016 Annual Reliability and Maintainability Symposium (RAMS), pp. 1–6. IEEE (2016). https://doi.org/10.1109/RAMS.2016.7447986
19. Volk, M., Junges, S., Katoen, J.-P.: Advancing dynamic fault tree analysis - get succinct state spaces fast and synthesise failure rates. In: Skavhaug, A., Guiochet, J., Bitsch, F. (eds.) SAFECOMP 2016. LNCS, vol. 9922, pp. 253–265. Springer, Cham (2016). https://doi.org/10.1007/978-3-319-45477-1_20
20. Yevkin, O.: An improved modular approach for dynamic fault tree analysis. In: Annual Reliability and Maintainability Symposium, pp. 1–5. IEEE (2011). https://doi.org/10.1109/RAMS.2011.5754437

Composition of Fault Forests

Danielle Stewart[1]([✉]), Michael Whalen[1], Mats Heimdahl[1], Jing (Janet) Liu[2],
and Darren Cofer[2]

[1] University of Minnesota, Minneapolis, MN, USA
{dkstewar,mwwhalen,heimdahl}@umn.edu
[2] Collins Aerospace – Applied Research & Technology, Cedar Rapids, IA, USA
{jing.liu,darren.cofer}@collins.com

Abstract. Safety analysis is used to ensure that critical systems operate within some level of safety when failures are present. As critical systems become more dependent on software components, it becomes more challenging for safety analysts to comprehensively enumerate all possible failure causation paths. Any automated analyses should be sound to sufficiently prove that the system operates within the designated level of safety. This paper presents a compositional approach to the generation of fault forests (sets of fault trees) and minimal cut sets. We use a behavioral fault model to explore how errors may lead to a failure condition. The analysis is performed per layer of the architecture and the results are automatically composed. A complete formalization is given. We implement this by leveraging minimal inductive validity cores produced by an infinite state model checker. This research provides a sound alternative to a monolithic framework. This enables safety analysts to get a comprehensive enumeration of all applicable fault combinations using a compositional approach while generating artifacts required for certification.

1 Introduction

Risk and safety analyses are important activities used to ensure that critical systems operate in an expected way. From nuclear power plants and airplanes to heart monitors and automobiles, critical systems are ubiquitous in our society. These systems are required to operate safely under nominal and faulty conditions. Proving that the system operates within some level of safety when failures are present is an important aspect of critical systems development and falls under the discipline of safety analysis. Safety analysis produces various safety related artifacts that are used during development and certification of critical systems [30]. Examples include *minimal cut sets* – each set represents the minimal set of faults that must all occur in order to violate a safety property and *fault trees* – the evaluation that determines all credible failure combinations which could cause an undesired top level hazard event. The fault tree can be transformed to an equivalent Boolean formula whose literals appear in the minimal cut sets. Since the introduction of minimal cut sets in the field of safety

© Springer Nature Switzerland AG 2021
I. Habli et al. (Eds.): SAFECOMP 2021, LNCS 12852, pp. 258–275, 2021.
https://doi.org/10.1007/978-3-030-83903-1_17

analysis, much research has been performed to address the generation of these sets and associated formulae [15,29,35]. As critical systems get larger, more minimal cut sets are possible with increasing cardinality. In recent years, symbolic model checking has been used to address scaling the analysis of systems with millions of minimal cut sets [4,10,31].

The state space explosion is a challenge when performing formal verification on industrial sized systems. This problem can arise from combining parallel processes together and attempting to reason monolithically over them. Compositional reasoning takes advantage of the hierarchical organizaton of a system model. A compositional approach verifies each component of the system in isolation and allows global properties to be inferred about the entire system [3]. The *assume-guarantee* paradigm is commonly used in compositional reasoning where the assumed behavior of the environment implies the guaranteed behavior of the component [13].

Using an assume-guarantee reasoning framework, we extend the definition of the nomimal transition system to allow for unconstrained guarantees. We use this idea to reason about all possible violations of a safety property per layer of analysis and then compose the results.

After we provide the formalization, we describe the implementation in the OSATE tool for the Architecture Analysis and Design Lanugage (AADL) [16]. AADL has two annexes that are of interest to us: the Assume-Guarantee Reasoning Environment (AGREE) [13] and the safety annex [32]. AGREE provides the assume-guarantee reasoning required for the transition system extension, and the safety annex allows us to define faults on component outputs. To implement the formalization, we look to recent work in formal verification. Ghassabani et al. developed an algorithm that traces a safety property to a minimal set of model elements necessary for proof; this is called the *all minimal inductive validity core* algorithm (All_MIVCs) [19,20]. Inductive validity cores produce the minimal sets of model elements necessary to prove a property. Each set contains the behavioral contracts – the requirement specifications of components – used in a proof. We collect all MIVCs per layer to generate the minimal cut sets and thus the fault trees to be composed.

This paper presents a compositional approach to generating *fault forests* (sets of fault trees) and associated minimal cut sets, allowing us to reason uniformly about faults in various types of system components and their impact on system properties. The main contributions of this research include the formalization of the composition of fault forests and its implementation, enabling safety analysts to get a comprehensive enumeration of all applicable fault combinations. The resulting fault trees correspond with the system architecture and reflect the interface specifications developed in the system under consideration. Our objective is to provide safety engineers with verification tools so that they do not lose sight of the fault forest for the trees.

The organization of the paper is as follows. Section 2 describes a running example, Sect. 3 outlines the formalization of this approach. The implementation is discussed in Sect. 4 and related work follows in Sect. 5. The paper ends with a conclusion and discussion of future work.

2   Running Example

In a typical Pressurized Water Reactor (PWR), the core inside of the reactor vessel produces heat. Pressurized water in the primary coolant loop carries the heat to the steam generator. Within the steam generator, heat from the primary coolant loop vaporizes the water in a secondary loop, producing steam. The steamline directs the steam to the main turbine causing it to turn the turbine generator, which in turn produces electricity. There are a few important factors that must be considered during safety assessment and system design. An unsafe climb in temperature can cause high pressure and hence pipe rupture, and high levels of radiation could indicate a leak of primary coolant. The following sensor system can be thought of as a simplified version of a subsystem within a PWR that monitors these factors. Each subsystem contain three sensors that monitor pressure, temperature, and radiation. If any of these conditions are too high, a shut down command is sent from the sensors to the parent components. The temperature, pressure, and radiation sensor subsystems each contain three associated sensors for redundancy. Each sensor reports the associated environmental condition to a majority voter component. If the majority of the sensors reports high, a shut down command is sent to the subsystem. If any subsystem reports a shut down command, the top level system will shut down. Pressure, radiation, and temperature all have associated thresholds for high values which we refer to as T_p, T_r, and T_t respectively. The safety properties P_i of interest in this system is: *if an environmental threshold is surpassed, then we shut down the system*. The specifications of these properties are shown at the top of Fig. 1.

For reference throughout this paper, we provide Fig. 1 which shows the guarantees and faults of interest for this running example. We do not show all guarantees and assumptions that are in the model, but only the ones of interest for the illustration.

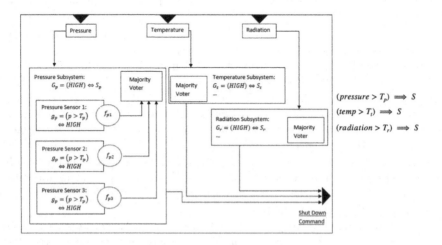

Fig. 1. Sensor system nominal and fault model details

3   Formalization

Given a state space U, a transition system (I, T) consists of an initial state predicate $I : U \to bool$ and a transition step predicate $T : U \times U \to bool$. We define the notion of reachability for (I, T) as the smallest predicate $R : U \to bool$ which satisfies the following formulas:

$$\forall u \in U. \; I(u) \Rightarrow R(u)$$
$$\forall u, u' \in U. \; R(u) \wedge T(u, u') \Rightarrow R(u')$$

A safety property $P : U \to bool$ is a state predicate. A safety property P holds on a transition system (I, T) if it holds on all reachable states, i.e., $\forall u. \; R(u) \Rightarrow P(u)$, written as $R \Rightarrow P$ for short. When this is the case, we write $(I, T) \vdash P$. We assume the transition relation has the structure of a top level conjunction. Given $T(u, u') = T_1(u, u') \wedge \cdots \wedge T_n(u, u')$ we will write $T = \wedge_{i=1..n} T_i$ for short. By further abuse of notation, T is identified with the set of its top-level conjuncts $\wedge_{i=1..n} T_i$. Thus, $T_i \in T$ means that T_i is a top-level conjunct of T, and $S \subseteq T$ means all top level conjuncts of S are top-level conjuncts of T.

The set of all nominal guarantees of the system G consists of conjunctive constraints $g \in G$. Given no faults (i.e., nominal system) and a transition relation T consisting of conjunctive constraints T_i, each g is one of the transition constraints T_i where:

$$T = g_1 \wedge g_2 \wedge \cdots \wedge g_n \tag{1}$$

We consider an arbitrary layer of analysis of the architecture and assume the property holds of the nominal relation $(I, T) \vdash P$. Let the set of all faults in the system be denoted as F. A fault $f \in F$ is a modification of the nominal constraint imposed by a guarantee. Without loss of generality, we associate a single fault and an associated fault probability with a guarantee. Each fault f_i is associated with an *activation literal*, af_i, that determines whether the fault is active or inactive. We extend the transition system so that we can view the system behavior in the presence of faults—or equivalently the absence of nominal constraints. To consider the system under the presence of faults, consider a set GF of modified guarantees in the presence of faults and let a mapping be defined from activation literals $af_i \in AF$ to these modified guarantees $gf_i \in GF$.

$$gf_i = if \, af_i \, then \, f_i \, else \, g_i$$

The transition system is composed of the set of modified guarantees GF and a set of conjunctions assigning each of the activation literals $af_i \in AF$ to false:

$$T' = gf_1 \wedge gf_2 \wedge \cdots \wedge gf_n \wedge \neg af_1 \wedge \neg af_2 \wedge \cdots \wedge \neg af_n \tag{2}$$

Theorem 1. *If $(I, T) \vdash P$ for T defined in Eq. 1, then $(I, T') \vdash P$ for T' defined in Eq. 2.*

Proof. By the mapping of each constrained activation literal $\neg af_i$ to the associated guarantee g_i and the constraint of the activation literals to be false, the result is immediate.   □

Consider the elements of T' as a set $GF \cup AF$, where GF are the potentially faulty guarantees and AF consists of the activation literals that determine whether a guarantee is faulty. This is a set that is considered by an SMT solver for satisfiability during the model checking engine procedures.

If the $af_i \in AF$ defined in T' are unconstrained, this allows more behaviors to the transition system and could cause a violation of P. If so, a counterexample may be produced. For each counterexample, we can partition AF into two sets that we call *non-faulty variables (NFV)* and *faulty variables (FV)*. The set NFV consists of a set of activation literals that are constrained to be false throughout the counterexample, and FV contains those that can be non-deterministically assigned any valuation at some point in the trace. By mapping some of the variables in AF to false, we know that their associated guarantees in GF are non-faulty for all considered executions. We define $T'(NFV)$ as a relaxation of T' (2):

$$T'(NFV) = gf_1 \wedge gf_2 \wedge \cdots \wedge gf_n \wedge \bigwedge \{\neg af_i \,|\, af_i \in NFV\}$$

The activation literals constrained to be false in $T'(NFV)$ indicate that their associated guarantees to be valid. In the remainder of this section, we assume that all $af_i \in AF$ are unconstrained and when given a true valuation will lead to a violation of the associated guarantee. This violation causes the output that the guarantee constrains to become non-deterministic. The Boolean variables in FV correspond to Boolean variables in the fault tree.

Definition 1. *A fault tree FT is a pair $(r, \mathcal{L})$ where:*

  r: the root r is a negated desirable property,
  $\mathcal{L}$: a Boolean equation whose literals are faulty variables.

All literals af of the Boolean equation $\mathcal{L}$ are elements of the set FV. A fault tree may correspond to a single layer of the system architecture where the root r is a violated guarantee or a violated safety property depending on the parent component under analysis. The tree may also describe the relationship between faults and multiple layers of the system architecture. The root r still corresponds to a violated guarantee or property, but the structure of the Boolean formula $\mathcal{L}$ will reflect the layers of the system architecture. If r is a violated safety property, then $r \in P$. If r is a violated guarantee for some lower level parent component, then $r \in \pi$, where π is the set of parent component guarantees.

Definition 2. *A fault tree $FT = (r, \mathcal{L})$ is valid if and only if a true valuation for r and for all $af \in \mathcal{L}$ is satisfiable given the respective transition system constraints.*

The hierarchy of the fault tree is dependent on the associated Boolean formula. A more intuitive structure is that of *disjunctive normal form* (DNF) as seen in both fault trees depicted in Fig. 2, but DNF is not required under our definition of a fault tree.

Traditionally, a safety property is a property of the system and in the assume-guarantee reasoning environment is a top level guarantee. In the following formalism, each layer of analysis is viewed as distinct from the system hierarchy as the proof is being constructed, and the properties we wish to prove are guarantees of a component. We use the notation P to refer to the set of all parent properties at a given layer of analysis. If the analysis is being performed at the top level, these are all safety properties of the system. If the analysis is being performed at an intermediate level, these are all guarantees of the parent component.

A goal of compositional safety analysis is to reflect failures of leaf and intermediate components at the top level. Not all guarantees must be valid to prove a parent level guarantee. To this end, we wish to make a distinction between all guarantees of a component and those that are required to prove parent guarantees. The subset π of P are the guarantees that must be valid to prove the guarantees of a parent component. These are the critical guarantees of a component. Given that there may be multiple safety properties and multiple intermediate level guarantees, we do not compose single fault trees per layer, but rather forests of trees.

Definition 3. *A fault forest FF is a set of fault trees.*

Definition 4. *A fault forest FF is valid if and only if for all $FT \in FF$, the fault tree FT is valid as per Definition 2.*

The goal of this formalization is to show that the composition of fault forests results in a valid fault forest. First, we assume we can derive all minimal counterexamples to the proof of a property (or guarantee) at any layer of compositional assume-guarantee analysis. Then we prove that after composition, the tree we obtain is a fault tree describing the system in the presence of faults. In Sect. 4, we discharge the assumption and show how we derive a valid fault forest for each layer of analysis. Since a fault forest is only valid with respect to the transition system from whence it came, we will now iteratively extend the model with each composition step.

Components and Their Composition: To prove each parent component guarantee $\pi_i \in \pi$, a certain subset of child guarantees are required to be non-faulty, i.e., the associated activation literals are given a false valuation. We use the set NFV to denote the non-faulty variables of the children components that are required to prove parent guarantees π. These non-faulty variables are used in the relaxation of T' (Eq. 2). This can be stated as $(I, T'(NFV)) \vdash \pi$.

The violation of certain child guarantees may lead to the violation of a parent guarantee π_i. The activation literals of the child are given a true valuation and are denoted as FV: faulty variables. A set of faulty variables of the children components contain the activation literals that correspond to leaves of a fault tree $\mathcal{L}$ with the root $r = \neg\pi_i$ for parent guarantee π_i. In other words, the fault tree $FT_i \in FF$ is associated with a property π_i. The non-faulty variables NFV contain the valid child guarantees that are required to prove π_i, and the fault tree FT_i reflects the child guarantee violations that may lead to the violation of π_i.

Definition 5. *A component is the tuple* $Comp(M, FF, NFV, \pi)$ *where:*

- *M: the model consisting of the set of all children properties P_c extended with non-deterministic faults: $gf_i \in P_c$ where $gf_i =$ if af_i then f_i else g_i,*
- *FF: the ordered set of fault trees for this component,*
- *NFV: the set of non-faulty variables, $NFV \subseteq P_c$,*
- *π: the ordered set of properties $\pi \subseteq P$ such that $(I, T'(NFV)) \vdash \pi$, i.e., all properties π hold if the variables in NFV are given a true valuation.*

and $FT_i \in FF$ corresponds to $\pi_i \in \pi$ for each of the i properties: the root of FT_i is $\neg\pi_i$.

Given the definition of a component, we now discuss what it means to compose components. Each layer of composition moves iteratively closer to a monolithic model by the enlargement of each set described in a component. To begin this iterative process, we define the composition of fault forests. To show that the composition of fault trees results in a valid fault tree, let ϕ be a function $\phi : B \times B \rightarrow B$ for Boolean equations B. We use this mapping to define the composition of parent component fault tree FT_p and child component fault tree FT_c, where $FT_c = (r_c, \mathcal{L}_c)$ and $FT_p = (r_p, \mathcal{L}_p)$.

$$FT_c \circ FT_p = \phi(FT_c, FT_p) = \begin{cases} (r_p, \mathcal{L}_p(r_c, \mathcal{L}_c)) & r_c \in \mathcal{L}_p \\ (r_p, \mathcal{L}_p) & r_c \notin \mathcal{L}_p \end{cases} \tag{3}$$

where $\mathcal{L}_p(r_c, \mathcal{L}_c)$ is the replacement of af_{r_c} in $\mathcal{L}_p$ with $(r_c, \mathcal{L}_c)$. Intuitively, each of the violated guarantees has an associated activation literal. If an activation literal is found in the parent leaf equation $\mathcal{L}_p$, replace that activation literal (af_{r_c}) with the associated violated child guarantee (r_c).

Let n be the number of properties for some parent component p and let m be the number of properties for some child component c. Then the parent fault forest FF_p is a mapping $FF_p : S_1 \rightarrow B$ for $S_1 = \{1, 2, \ldots, m\}$ and the set of Boolean equations B and $FF_c : S_2 \rightarrow B$ for $S_2 = \{1, 2, \ldots n\}$. And let ϕ_F be a function $\phi_F : seq(B) \times seq(B) \rightarrow seq(B)$ for finite sequences of Boolean equations $seq(B)$. We use this function to define the composition of parent and child component fault forests $FF_p = \{(r_{p1}, \mathcal{L}_{p1}), \ldots, (r_{pm}, \mathcal{L}_{pm})\}$ and $FF_c = \{(r_{c1}, \mathcal{L}_{c1}), \ldots, (r_{cn}, \mathcal{L}_{cn})\}$. ϕ_F is a mapping such that for all $i \in S_1$ and for all $j \in S_2$:

$$FF_c \circ FF_p = \phi_F(FF_c, FF_p) = \begin{cases} (r_{pi}, \mathcal{L}_{pi}(r_{cj}, \mathcal{L}_{cj})) & r_{cj} \in \mathcal{L}_{pi} \\ (r_{pi}, \mathcal{L}_{pi}) & r_{cj} \notin \mathcal{L}_{pi} \end{cases} \tag{4}$$

where $\mathcal{L}_{pi}(r_{cj}, \mathcal{L}_{cj})$ is the replacement of $af_{r_{cj}}$ in $\mathcal{L}_{pi}$ with $(r_{cj}, \mathcal{L}_{cj})$.

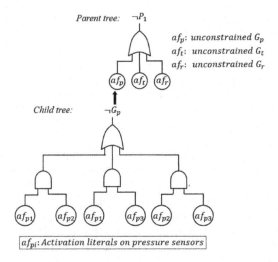

Fig. 2. Sensor system composition of fault trees

Each literal in the formula $\mathcal{L}_p$ is a fault activation literal af_i. If af_i has its associated guarantee gf_i in the set of child roots r_c, then the mapping ϕ_F will extend af_i in $\mathcal{L}_p$ with the leaf formula of the child root gf_i. The resulting fault forest is a sequence of fault trees $FF = \{(r_{pk}, \mathcal{L}_k) : k = 1, \ldots, m\}$. The roots of the resulting forest are the same roots as the parent forest while the leaf formulae may change based on replacement.

We return to the sensor system example to illustrate this mapping. Graphically, this is represented in Fig. 2. The top level (parent) component is defined as: $Comp_p(M_p, FF_p, NFV_p, \pi_p)$ and $FF_p = \{(\neg P, af_p \vee af_t \vee af_r)\}$ where each activation literal is associated with the unconstrained guarantees G_p, G_t, and G_r. The child layer has a fault forest consisting of three fault trees, one for each subsystem. The pressure subsystem fault tree is $FT_p = (\neg G_p, (af_{p1} \wedge af_{p2}) \vee (af_{p1} \wedge af_{p3}) \vee (af_{p2} \wedge af_{p3}))$. The leaf formulae for each subsystem tree corresponds to pairwise combinations of active sensor faults. We now show the composition of the pressure subsystem child and top level parent fault trees.

The mapping ϕ_F iterates through each tree in the parent forest – in this case, we have only one. Then for each parent tree it iterates through the Boolean literals in $\mathcal{L}$. If there is a match between a child root and a parent leaf, the replacement is made. We represent the unconstrained (violated) guarantee as $\neg G_p$ and it is associated with the fault activation literal af_p. Thus, af_p will be extended with $\{\neg G_p, (af_{p1} \wedge af_{p2}) \vee (af_{p1} \wedge af_{p3}) \vee (af_{p2} \wedge af_{p3})\}$. This extension is done for each leaf formula in $\mathcal{L}_p$ from the parent fault forest. The end result of the replacement is easy to see in Fig. 2.

We have provided the foundational definitions necessary to discuss what it means to compose components. The composition of child component $Comp_c$ and parent component $Comp_p$ is defined as:

Definition 6. $Comp_c(M_c, FF_c, NFV_c, \pi_c) \circ Comp_p(M_p, FT_p, NFV_p, \pi_p)$
$= Comp_\circ(M', FF', NFV', \pi')$ *where:*

- $M' = M_c \cup M_p$ *is the iterative enlargement of the model by combining children guarantees with parent guarantees,*
- $FF_c \circ FF_p$ *is the composed fault forest,*
- $NFV' = NFV_c \cup NFV_p$ *is the set of non-faulty variables,*
- $\pi' = \pi_c \cup \pi_p$ *are valid properties such that* $(I, T'(NFV')) \vdash \pi'$.

The enlargement of the model, M', iteratively flattens the composed layers by taking the union of children guarantees and parent guarantees. The fault forests are composed into a set of fault trees describing the enlarged model. The non-faulty variables from child and parent are combined into a set NFV' such that $(I, T'(NFV')) \vdash \pi'$. Given that in child and parent components, the properties π can be derived from the non-faulty variables, we show that this relationship holds after composition. To state $(I, T'(NFV)) \vdash \pi$, we use the shorthand $NFV \vdash \pi$.

Theorem 2. *If* $NFV_c \vdash \pi_c$ *and* $NFV_p \vdash \pi_p$, *then* $NFV' \vdash \pi'$

Proof. Assume antecedent. Let $p' \in \pi'$. If $p' \in \pi_c$ then $NFV_c \vdash p'$ and likewise if $p' \in \pi_p$, then $NFV_p \vdash p'$. In either case, $NFV_c \cup NFV_p = NFV' \vdash \pi'$. $\quad\square$

Composition of Fault Trees and Forests: We work under the *monotonicity assumption*, commonly adopted in safety analysis, that an additional fault cannot cancel the effect of existing faults. Without this assumption, we cannot show that the resulting fault tree is valid. Given Definition 2, we show that the composition of two fault trees results in a valid fault tree. We will then extend this to show that the composition of two fault forests results in a valid fault forest.

Lemma 1. *If* FT_c *and* FT_p *are valid fault trees, then their composition* $\phi(FT_c, FT_p)$ *is also a valid fault tree.*

Proof. Assume the antecedent. Then $(r_c, \mathcal{L}_c)$ is satisfiable with regard to the child component transition system and all $af \in \mathcal{L}_c$ and r_c are given true valuations.

Case 1: If the child root $\neg g_i$ does not have an associated $af_i \in \mathcal{L}_p$, then $\phi(FT_c, FT_p) = FT_p$ and the inclusion of the additional constraints from the child transition system in M_c does not negate the effects of the faults in FT_p. Thus, it is a valid fault tree.

Case 2: If the child root $\neg g_i$ has an associated $af_i \in \mathcal{L}_p$, then af_i has a true valuation. Given the mapping defined between guarantees and activation literals, replacement of $af_i \in \mathcal{L}_p$ with $\neg g_i$ preserves satisfiability. Furthermore, by the monotonicity assumption, the addition of more constraints ($af \in \mathcal{L}_c$) to the Boolean formula does not change satisfiability in the extended transition system.

In all cases, $\phi(FT_c, FT_p)$ is a valid fault tree. $\quad\square$

Lemma 2. *If FF_c and FF_p are valid fault forests, then their composition $\phi(FF_c, FF_p)$ is also a valid fault forest.*

Proof. Assume the antecedent. Then for all $FT_j \in FF_p$ and $FT_i \in FF_c$, FT_i and FT_j are valid fault trees as per Definition 4. For each iteration defined in the mapping ϕ_F, apply Lemma 1 and the monotonicity assumption. □

We have shown that a single layer of composition produces valid fault forests. To perform this analysis across n layers of architecture we use induction to show that the resulting fault forest is valid. The notation ϕ_F^n indicates the iterated function ϕ_F which is a successive application of ϕ_F with itself n times. Assume the fault forest FF_0 is obtained at the leaf level of the architecture.

Theorem 3. *If $\phi_F^n(FF_{n-1}, FF_n)$ is a valid fault forest, then $\phi^{n+1}(FF_n, FF_{n+1})$ is a valid fault forest.*

Proof. Base case: Each fault forest per layer is valid by construction. By Lemma 2, $\phi_F(FF_0, FF_1)$ is a valid fault forest.

Inductive assumption: Assume $\phi_F^n(FF_{n-1}, FF_n)$ is a valid fault forest.

$$\phi_F^{n+1}(FF_n, FF_{n+1}) = ((FF_0 \circ FF_1) \circ FF_2) \circ \cdots \circ FF_n) \circ FF_{n+1})$$
$$= \phi_F^n(FF_{n-1}, FF_n) \circ FF_{n+1}$$

By inductive assumption and Lemma 2, $\phi_F^{n+1}(FF_n, FF_{n+1})$ is a valid fault forest. □

After applying these techniques to the pressurized water reactor example, the resulting fault forest consists of one tree associated with the single top level safety hazard as shown in Fig. 3.

In this section, we have formalized the idea that fault trees (and forests) can be composed without losing the validity of each composed tree. We proved that this can be performed iteratively across an arbitrary number of layers.

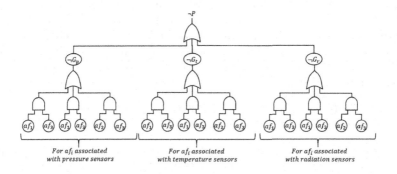

Fig. 3. Sensor system fault forest

4   Implementation

To implement the formalism described in Sect. 3, we must compute minimal cut sets per layer of analysis, transform them into their related Boolean formula, and compose them. As previously described, Ghassabani et al. developed the *all minimal inductive validity core* algorithm (All_MIVCs) [19,20]. The All_MIVCs algorithm gives the minimal set of contracts required for proof of a safety property. If all of these sets are obtained, we have insight into every proof for the property. Thus, if we violate at least one contract from every MIVC set, we have in essence "broken" every proof. The idea is that the hitting sets of all MIVCs produces the minimal cut sets.

4.1   Formal Background

JKind is an open-source industrial infinite-state inductive model checker for safety properties [17]. Models and properties in JKind are specified in Lustre [22], a synchronous dataflow language, using the theories of linear real and integer arithmetic. JKind uses SMT-solvers to prove and falsify multiple properties in parallel.

Each step of induction is sent to an SMT (Satisfiabilty Modulo Theory)-solver to check for *satisfiability*, i.e. there exists a total truth assignment to a given formula that evaluates to true. If there does not exist such an assignment, the formula is considered *unsatisfiable*. A k-induction model checker utilizes parallel SMT-solving engines at each induction step to glean information about the proof of a safety property. The transition formula is translated into clauses such that satisfiability is preserved. Expression of the base and induction steps of a temporal induction proof as SAT problems is straightforward and is shown below for step k:

$$I(s_0) \wedge T(s_0, s_1) \wedge \cdots \wedge T(s_{k-1}, s_k) \wedge \neg P(s_k)$$

When proving correctness it is shown that the formulas are *unsatisfiable*, i.e., the property P is provable. The idea behind finding an *inductive validity core* (IVC) for a given property P is based on inductive proof methods used in SMT-based model checking, such as k-induction and IC3/PDR [23]. Generally, an IVC computation technique aims to determine, for any subset $S \subseteq T$, whether P is provable by S. A minimal subset that satisfies P is seen as a minimal proof explanation and called a minimal inductive validity core.

Definition 7. *Inductive Validity Core (IVC) [19]:* $S \subseteq T$ *for* $(I, T) \vdash P$ *is an Inductive Validity Core, denoted by* $IVC(P, S)$, *iff* $(I, S) \vdash P$.

Definition 8. *Minimal Inductive Validity Core (MIVC) [20]:* $S \subseteq T$ *is a minimal Inductive Validity Core, denoted by* $MIVC(P, S)$, *iff* $IVC(P, S) \wedge \forall T_i \in S$. $(I, S \setminus \{T_i\}) \nvdash P$.

The *constraint system* consists of the constrained formulas of the transition system and the negation of the property. The All_MIVCs algorithm collects all *minimal unsatisfiable subsets* (MUSs) of a constraint system generated from a transition system at each induction step [2, 20].

Definition 9. *A Minimal Unsatisfiable Subset (MUS) M of a constraint system C is a set $M \subseteq C$ such that M is unsatisfiable and $\forall c \in M : M \backslash \{c\}$ is satisfiable.*

The MUSs are the minimal explanation of the infeasibility of this constraint system; equivalently, these are the minimal sets of model elements necessary for proof of the safety property.

Returning to our running example, this can be illustrated by the following. Given the constraint system $C = \{G_p, G_t, G_r, \neg P\}$, a minimal explanation of the infeasability of this system is the set $\{G_p, G_t, G_r, \}$. If all three guarantees hold, then P (the disjunction of these guarantees) is provable.

In the case of an UNSAT system, we may ask: what will correct this unsatisfiability? A related set answers this question:

Definition 10. *A Minimal Correction Set (MCS) M of a constraint system C is a subset $M \subseteq C$ such that $C \setminus M$ is satisfiable and $\forall M' \subset M : C \setminus M'$ is unsatisfiable.*

An MCS can be seen to "correct" the infeasability of the constraint system by the removal from C the constraints found in an MCS. Returning to the PWR example, the MCSs of the constraint system C are $MCS_1 = \{G_t\}$, $MCS_2 = \{G_p\}$, $MCS_3 = \{G_r\}$. If any single guarantee is violated, a shut down from that subsystem may not get sent when it should and the safety property P will be violated. This corresponds exactly to the definition of a minimal cut set.

For the following definitions, we remind readers of the extended transition system defined in Eq. 2 of Sect. 3 and that the elements of T' are the set $GF \cup AF$ for potentially faulty guarantees GF and activation literals AF. We use the notation $af \rightarrow \{true, false\}$ to indicate a constraint on the literal af.

Definition 11. *Given a constraint system C, a cut set S of a top level event $\neg P$ is a set $S \subseteq AF \subseteq C$ such that $\forall af \in S$, $af \rightarrow \{true\}$ and $S \cup \{\neg P\}$ is satisfiable in C.*

Intuitively, a cut set is a true valuation for some subset of fault activation literals within a constraint system containing such that the constraint system is satisfiable given those true valuations and the violation of a safety property.

Definition 12. *A cut set S is minimal if and only if $\forall af \in S$, $S \setminus \{af\} \cup \{\neg P\}$ is unsatisfiable.*

Our approach in computing minimal cut sets through the use of inductive validity cores is to supply activation literals constrained to be false to the algorithm. The resulting MCSs consist of elements $\neg af_i$. The removal of this constraint from the constraint system results in non-deterministically true activation literals. By the definition of an MCS, we know that $C \setminus MCS$ is satisifiable.

This removal of constraints from C removes the *false* constraint from each element in the MCS. Liffiton et al. showed that any subset of a satisfiable set is also satisfiable [24], so we know that for set S consisting of elements of MCS with constraints removed, $S \cup \{\neg P\}$ is also satisfiable. This is the definition of a cut set. Minimality comes directly from the definition of a minimal correction set.

A duality exists between the MUSs of a constraint system and the MCSs as established by Reiter [28]. This duality is defined in terms of *Minimal Hitting Sets (MHS)*.

Definition 13. *A hitting set of a collection of sets A is a set H such that every set in A is "hit" by H; H contains at least one element from every set in A.*

Every MUS of a constraint system is a minimal hitting set of the system's MCSs, and likewise every MCS is a minimal hitting set of the system's MUSs. This is noted in previous work [14,24] and the proof of such is given by Reiter (Theorem 4.4 and Corollary 4.5) [28].

4.2   Algorithm Implementation

The algorithms in this paper are implemented in the Safety Annex [32] for the Architecture Analysis and Design Language (AADL) [1] and require the Assume-Guarantee Reasoning Environment (AGREE) [13] to annotate the AADL model in order to perform verification using the back-end model checker JKind [17]. For more information on the application of the safety annex in practice, see previous work [32–34].

In the formalism, any guarantee in the model had an associated fault activation literal and could be unconstrained. In the implementation, we rely on the fault model created in the safety annex to dictate which output constraints are modified (i.e., which guarantees can be violated) and how they are modified. A user may define multiple, single, or no faults on a single output. Each explicit fault defined in the safety annex is added to the Lustre program as are assocated fault activation literals [32,34]. This corresponds to the f_i and af_i described in Sect. 3.

The All_MIVCs algorithm requires specific equations in the Lustre model to be flagged for consideration in the analysis; these we call *IVC algorithm elements*. All equations in the model can be used as IVC algorithm elements or one can specify directly the equations to consider. In this implementation, the IVC algorithm elements are added differently depending on the layer. In the leaf architectural level, fault activation literals are added to the IVC algorithm elements and are constrained to *false*. In middle or top layers, supporting guarantees are added. This is shown in Fig. 4. The figure shows an arbitrary architecture with two analysis layers: top and leaf. The top layer analysis adds G as IVC algorithm element; the leaf layer analysis adds f_1 and f_2.

A requirement of the hitting set algorithm is that to find all MCSs, all MUSs must be known. Ghassabani et al. [20] showed that finding all MIVCs is as hard as model checking. Once the MIVC analysis is complete for a property at a

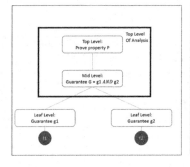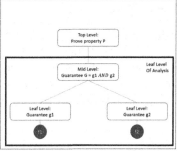

Fig. 4. Illustration of two layers of analysis

given layer, a hitting set algorithm is used to generate the related MCSs [18]. Depending on the layer of analysis, the MCSs contain either guarantees (mid layer) or fault activation literals (leaf layer).

Algorithm 1: Compose Results

---

1   $R \leftarrow \texttt{All\_MCSs}(P) = \vee_{i=1}^{n} MCS_i$

2   where $MCS_i = \wedge_{j=1}^{m} gf_j$

3   **Function** resolve(R):

4   **for** $\forall$ *OR-node in* R **do**

5   **for** $\forall gf_j$ *in OR-node* **do**

6   **if** $\exists MCS(gf_j)$ **then**

7               $R \leftarrow$ replace gf_j in R with $\texttt{All\_MCSs}(gf_j)$;

8   resolve ($\texttt{All\_MCSs}(gf_j)$);

9   **else**

10               $R \leftarrow$ replace gf_j in R with af_j;

11   convert R to DNF

The composition of these results is performed top down and shown in Algorithm 1. For each guarantee found in an MCS, a replacement is made with the guarantee's own MCSs. This is done recursively until all replacements have been made (line 7, 8 of Algorithm 1). If on the other hand there are no MCSs for a given guarantee, that guarantee is replaced by its associated fault activation literal (line 10). At the leaf level of analysis, no guarantees have associated MCSs (there are no children properties) and thus reaches the end of recursion. At that time, the formula is converted back into disjunctive normal form of fault activation literals to finish the translation into the traditional fault tree (line 11). The fault tree that is produced has a depth associated with the architecture of the system model. The gates supported in this tool include *And* and *Or* gates.

Theorem 4. *Algorithm 1 terminates*

Proof. No infinite sets are generated by the All_MIVCs or minimal hitting set algorithms [20,25]; therefore, for all g_i in the model, All_MCSs(g_i) is a finite set and $MCS(g_i)$ is a finite set. Each call to Resolve processes a guarantee that was not previously resolved, and for all g_i at the leaf layer of analysis, All_MCSs($(g_i) = \emptyset$. Given that there are finite layers in a model, the algorithm terminates.   □

5   Related Work

Minimal cut sets generated by monolithic analysis look at explicitly defined faults throughout the architecture and attempt through various techniques to find the minimal violating set for a particular property. We now outline some of the common monolithic approaches to minimal cut set generation.

The representation of Boolean formulae as Binary Decision Diagrams (BDDs) was first formalized in the mid 1980s [11] and was extended to the representation of fault trees not many years later [27]. After this formalization, the BDD approach to FTA provided a new approach to safety analysis. The model is constructed using a BDD, then a second BDD - usually slightly restructured - is used to encode minimal cut sets. Unfortunately, due to the structure of BDDs, the worst case is exponential in size in terms of the number of variables. In industrial sized systems, this is not realistically useful.

SAT based computation was introduced to address scalability problems in the BDD approach; initially it was used as a preprocessing step to simplify the decision diagram [8], but later was extended to allow for all minimal cut set processing and generation without the use of BDDs [7]. Since then, much research has focused on leveraging the power of model checking in the problems of safety assessment, e.g., [4,31,32]. Bozzano et al. formulated a Bounded Model Checking (BMC) approach to the problem by successively approximating the cut set generation and computations to allow for an "anytime approximation" in cases when the cut sets were simply too large and numerous to find [7]. These algorithms are implemented in xSAP [5] and COMPASS [6]. Another related work is contract based safety analysis performed using the OCRA tool [9]. By contrast, this research performs the minimal cut set computations in a purely compositional fashion.

The model based safety assessment tool AltaRica 3.0 [26] performs a series of processing to transform the model into a reachability graph and then compile to Boolean formula in order to compute the minimal cut sets. Other tools such as HiP-HOPS [12] have implemented algorithms that follow the failure propagations in the model and collect information about safety related dependencies and hazards. The Safety Analysis Modeling Language (SAML) [21] provides a safety specific modeling language that can be translated into a number of input languages for model checkers in order to provide model checking support for minimal cut set generation.

To our knowledge, a fully compositional approach to generating fault forests or minimal cut sets has not been introduced.

6 Conclusion and Future Work

We presented a formalism that defines the composition of fault forests by extending the transition system to allow for fault activation literals. This formalism is implemented by leveraging recent research in model checking techniques. Using the idea of minimal inductive validity cores (MIVCs), which are the minimal model elements necessary for a proof of a safety property, we are able to provide fault activation literals as model elements to the All_MIVCs algorithm which provides all the MIVCs that pertain to this property. These are used to generate minimal cut sets. Future work includes leveraging the system information embedded in this approach to generate graphical hierarchical fault trees as well as perform scalability studies that compare this approach with other non-compositional approaches to minimal cut set generation.

Acknowledgments. This research was funded by NASA contract NNL16AB07T and the University of Minnesota College of Science and Engineering Graduate Fellowship.

References

1. AS5506C: Architecture Analysis & Design Language (AADL), January 2017
2. Bendík, J., Ghassabani, E., Whalen, M., Černá, I.: Online enumeration of all minimal inductive validity cores. In: Johnsen, E.B., Schaefer, I. (eds.) SEFM 2018. LNCS, vol. 10886, pp. 189–204. Springer, Cham (2018). https://doi.org/10.1007/978-3-319-92970-5_12
3. Berezin, S., Campos, S., Clarke, E.M.: Compositional reasoning in model checking. In: de Roever, W.-P., Langmaack, H., Pnueli, A. (eds.) COMPOS 1997. LNCS, vol. 1536, pp. 81–102. Springer, Heidelberg (1998). https://doi.org/10.1007/3-540-49213-5_4
4. Bieber, P., Castel, C., Seguin, C.: Combination of fault tree analysis and model checking for safety assessment of complex system. In: Bondavalli, A., Thevenod-Fosse, P. (eds.) EDCC 2002. LNCS, vol. 2485, pp. 19–31. Springer, Heidelberg (2002). https://doi.org/10.1007/3-540-36080-8_3
5. Bittner, B., et al.: The xSAP safety analysis platform. In: TACAS (2016)
6. Bozzano, M., Bruintjes, H., Cimatti, A., Katoen, J.P., Noll, T., Tonetta, S.: The COMPASS 3.0 toolset. In: IMBSA 2017 (2017)
7. Bozzano, M., Cimatti, A., Griggio, A., Mattarei, C.: Efficient anytime techniques for model-based safety analysis. In: Kroening, D., Păsăreanu, C.S. (eds.) CAV 2015. LNCS, vol. 9206, pp. 603–621. Springer, Cham (2015). https://doi.org/10.1007/978-3-319-21690-4_41
8. Bozzano, M., et al.: Safety assessment of AltaRica models via symbolic model checking. Sci. Comput. Program. **98**, 464–483 (2015)
9. Bozzano, M., Cimatti, A., Mattarei, C., Tonetta, S.: Formal safety assessment via contract-based design. In: Automated Technology for Verification and Analysis (2014)

10. Bozzano, M., Cimatti, A., Tapparo, F.: Symbolic fault tree analysis for reactive systems. In: ATVA (2007)
11. Bryant, R.E.: Graph-based algorithms for Boolean function manipulation. Comput. IEEE Trans. **100**(8), 677–691 (1986)
12. Chen, D., Mahmud, N., Walker, M., Feng, L., Lönn, H., Papadopoulos, Y.: Systems modeling with EAST-ADL for fault tree analysis through HiP-HOPS*. IFAC Proc. Vol. **46**(22), 91–96 (2013)
13. Cofer, D., Gacek, A., Miller, S., Whalen, M.W., LaValley, B., Sha, L.: Compositional verification of architectural models. In: Goodloe, A.E., Person, S. (eds.) NFM 2012. LNCS, vol. 7226, pp. 126–140. Springer, Heidelberg (2012). https://doi.org/10.1007/978-3-642-28891-3_13
14. De Kleer, J., Williams, B.C.: Diagnosing multiple faults. Artif. Intell. **32**(1), 97–130 (1987)
15. Ericson, C.: Fault tree analysis - a history. In: Proceedings of the 17th International Systems Safety Conference (1999)
16. Feiler, P., Gluch, D.: Model-Based Engineering with AADL: An Introduction to the SAE Architecture Analysis & Design Language. Addison-Wesley Professional, Boston (2012)
17. Gacek, A., Backes, J., Whalen, M., Wagner, L., Ghassabani, E.: The JKIND model checker. In: Chockler, H., Weissenbacher, G. (eds.) CAV 2018. LNCS, vol. 10982, pp. 20–27. Springer, Cham (2018). https://doi.org/10.1007/978-3-319-96142-2_3
18. Gainer-Dewar, A., Vera-Licona, P.: The minimal hitting set generation problem: algorithms and computation. SIAM J. Discrete Math. **31**(1), 63–100 (2017)
19. Ghassabani, E., Gacek, A., Whalen, M.W.: Efficient generation of inductive validity cores for safety properties. CoRR abs/1603.04276 (2016). http://arxiv.org/abs/1603.04276
20. Ghassabani, E., Whalen, M.W., Gacek, A.: Efficient generation of all minimal inductive validity cores. In: 2017 Formal Methods in Computer Aided Design (FMCAD), pp. 31–38 (2017)
21. Gudemann, M., Ortmeier, F.: A framework for qualitative and quantitative formal model-based safety analysis. In: HASE 2010 (2010)
22. Halbwachs, N., Caspi, P., Raymond, P., Pilaud, D.: The synchronous dataflow programming language lustre. IEEE **79**(9), 1305–1320 (1991)
23. Kahsai, T., Garoche, P.-L., Tinelli, C., Whalen, M.: Incremental verification with mode variable invariants in state machines. In: Goodloe, A.E., Person, S. (eds.) NFM 2012. LNCS, vol. 7226, pp. 388–402. Springer, Heidelberg (2012). https://doi.org/10.1007/978-3-642-28891-3_35
24. Liffiton, M.H., Previti, A., Malik, A., Marques-Silva, J.: Fast, flexible MUS enumeration. Constraints **21**(2), 223–250 (2015). https://doi.org/10.1007/s10601-015-9183-0
25. Murakami, K., Uno, T.: Efficient algorithms for dualizing large-scale hypergraphs. In: 2013 Proceedings of the Fifteenth Workshop on Algorithm Engineering and Experiments (ALENEX). SIAM (2013)
26. Prosvirnova, T.: AltaRica 3.0: a Model-based approach for safety analyses. Theses, Ecole Polytechnique, November 2014. https://pastel.archives-ouvertes.fr/tel-01119730
27. Rauzy, A.: New algorithms for fault trees analysis. Reliab. Eng. Syst. Saf. **40**(3), 203–211 (1993)
28. Reiter, R.: A theory of diagnosis from first principles. Artif. Intell. **32**(1), 57–95 (1987)

29. Ruijters, E., Stoelinga, M.: Fault tree analysis: a survey of the state-of-the-art in modeling, analysis and tools. Comput. Sci. Rev. **15–16**, 29–62 (2015)
30. SAE ARP4754A: Guidelines for Development of Civil Aircraft and Systems, December 2010
31. Schäfer, A.: Combining real-time model-checking and fault tree analysis. In: Araki, K., Gnesi, S., Mandrioli, D. (eds.) FME 2003. LNCS, vol. 2805, pp. 522–541. Springer, Heidelberg (2003). https://doi.org/10.1007/978-3-540-45236-2_29
32. Stewart, D., Liu, J., Heimdahl, M., Whalen, M., Cofer, D., Peterson, M.: The safety annex for architecture analysis and design language. In: 10th Edition European Congress Embedded Real Time Systems, January 2020
33. Stewart, D., Liu, J.J., Cofer, D., Heimdahl, M., Whalen, M.W., Peterson, M.: Aadl-based safety analysis using formal methods applied to aircraft digital systems. Reliab. Eng. Syst. Saf. **213**, 107649 (2021). https://doi.org/10.1016/j.ress.2021.107649, https://www.sciencedirect.com/science/article/pii/S0951832021001903
34. Stewart, D., Whalen, M.W., Cofer, D., Heimdahl, M.P.E.: Architectural modeling and analysis for safety engineering. In: Bozzano, M., Papadopoulos, Y. (eds.) IMBSA 2017. LNCS, vol. 10437, pp. 97–111. Springer, Cham (2017). https://doi.org/10.1007/978-3-319-64119-5_7
35. Vesely, W., Goldberg, F., Roberts, N., Haasl, D.: Fault tree handbook. Technical Report, US Nuclear Regulatory Commission (1981)

Author Index

Printed in the United States
by Baker & Taylor Publisher Services